高等职业教育系列教材

自动化专业英语
第 2 版

主　编　徐存善　毛琰虹　党菲菲
副主编　周志宇　黄　靓　朱　理
参　编　蒋志豪　余梦媛　王　颖

机械工业出版社

本书由4部分组成,即电力电子技术基础篇、楼宇智能化技术篇、电气自动控制技术和工业4.0与中国品牌制造技术篇。本书内容基本涵盖了电力电子技术基础、仪器仪表使用与维护、电能输送、电力系统监控、继电器工作原理、楼宇智能化、办公自动化、可编程逻辑控制器、传感器技术、自动化控制技术、第四次工业革命和具有代表性的中国品牌制造技术等。本书内容丰富、题材广泛、语言通俗易懂,能满足不同层次的学习对象对专业英语的学习需求。本书共有22个单元,每个单元包括课文、生词、专业术语、长难句解析、翻译技巧和阅读材料。在本书的最后4个单元的实用英语中,分别用相当篇幅介绍了怎样阅读英文招聘广告、怎样用英文写个人简历、如何写英文求职信等应用文体,以及英语面试过程中的常用技巧,目的是为了使毕业生在就业竞争中能胜人一筹。

本书适合高职高专院校电气自动化、生产过程自动化、电力系统自动化和电气控制技术类专业的学生使用,还可供相关专业的技术人员参考学习。

本书配有授课电子课件,需要的教师可登录 www.cmpedu.com 免费注册,审核通过后下载,或联系编辑索取(QQ:1239258369,电话:010-88379739)。

图书在版编目(CIP)数据

自动化专业英语/徐存善,毛琰虹,党菲菲主编. —2版. —北京:机械工业出版社,2017.8(2024.7重印)
高等职业教育系列教材
ISBN 978-7-111-57624-2

Ⅰ.①自… Ⅱ.①徐… ②毛… ③党… Ⅲ.①自动化技术—英语 Ⅳ.①TP1

中国版本图书馆 CIP 数据核字(2017)第 188177 号

机械工业出版社(北京市百万庄大街22号 邮政编码100037)
策划编辑:王　颖　责任编辑:李文轶
责任校对:王　延　责任印制:张　博
北京建宏印刷有限公司印刷
2024年7月第2版第8次印刷
184mm×260mm・14.75印张・353千字
标准书号:ISBN 978-7-111-57624-2
定价:39.90元

电话服务　　　　　　　　　网络服务
客服电话:010-88361066　　机 工 官 网:www.cmpbook.com
　　　　　010-88379833　　机 工 官 博:weibo.com/cmp1952
　　　　　010-68326294　　金　书　网:www.golden-book.com
封底无防伪标均为盗版　　　机工教育服务网:www.cmpedu.com

高等职业教育系列教材
电子类专业编委会成员名单

主　　任　曹建林

副 主 任　张中洲　张福强　董维佳　俞　宁　杨元挺　任德齐
　　　　　　华永平　吴元凯　蒋蒙安　梁永生　曹　毅　程远东
　　　　　　吴雪纯

委　　员　（按姓氏笔画排序）
　　　　　　于宝明　王卫兵　王树忠　王新新　牛百齐　吉雪峰
　　　　　　朱小祥　庄海军　刘　松　刘　勇　孙　刚　孙　萍
　　　　　　孙学耕　李菊芳　杨打生　杨国华　何丽梅　邹洪芬
　　　　　　汪赵强　张静之　陈子聪　陈东群　陈必群　陈晓文
　　　　　　邵　瑛　季顺宁　赵新宽　胡克满　姚建永　聂开俊
　　　　　　贾正松　夏西泉　高　波　高　健　郭　兵　郭　勇
　　　　　　郭雄艺　黄永定　章大钧　彭　勇　董春利　程智宾
　　　　　　曾晓宏　詹新生　蔡建军　谭克清　戴红霞

秘 书 长　胡毓坚

出 版 说 明

《国务院关于加快发展现代职业教育的决定》指出：到2020年，形成适应发展需求、产教深度融合、中职高职衔接、职业教育与普通教育相互沟通，体现终身教育理念，具有中国特色、世界水平的现代职业教育体系，推进人才培养模式创新，坚持校企合作、工学结合，强化教学、学习、实训相融合的教育教学活动，推行项目教学、案例教学、工作过程导向教学等教学模式，引导社会力量参与教学过程，共同开发课程和教材等教育资源。机械工业出版社组织国内80余所职业院校（其中大部分是示范性院校和骨干院校）的骨干教师共同规划、编写并出版的"高等职业教育系列教材"，已历经十余年的积淀和发展，今后将更加紧密结合国家职业教育文件精神，致力于建设符合现代职业教育教学需求的教材体系，打造充分适应现代职业教育教学模式的、体现工学结合特点的新型精品化教材。

在本系列教材策划和编写的过程中，主编院校通过编委会平台充分调研相关院校的专业课程体系，认真讨论课程教学大纲，积极听取相关专家意见，并融合教学中的实践经验，吸收职业教育改革成果，寻求企业合作，针对不同的课程性质采取差异化的编写策略。其中，核心基础课程的教材在保持扎实的理论基础的同时，增加实训和习题以及相关的多媒体配套资源；实践性课程的教材则强调理论与实训紧密结合，采用理实一体的编写模式；实用技术型课程的教材则在其中引入了最新的知识、技术、工艺和方法，同时重视企业参与，吸纳来自企业的真实案例。此外，根据实际教学的需要对部分内容进行了整合和优化。

归纳起来，本系列教材具有以下特点：

1）围绕培养学生的职业技能这条主线来设计教材的结构、内容和形式。

2）合理安排基础知识和实践知识的比例。基础知识以"必需、够用"为度，强调专业技术应用能力的训练，适当增加实训环节。

3）符合高职学生的学习特点和认知规律。对基本理论和方法的论述容易理解、清晰简洁，多用图表来表达信息；增加相关技术在生产中的应用实例，引导学生主动学习。

4）教材内容紧随技术和经济的发展而更新，及时将新知识、新技术、新工艺和新案例等引入教材。同时注重吸收最新的教学理念，并积极支持新专业的教材建设。

5）注重立体化教材建设。通过主教材、电子教案、配套素材光盘、实训指导和习题及解答等教学资源的有机结合，提高教学服务水平，为高素质技能型人才的培养创造良好的条件。

由于我国高等职业教育改革和发展的速度很快，加之我们的水平和经验有限，因此在教材的编写和出版过程中难免出现疏漏。我们恳请使用这套教材的师生及时向我们反馈质量信息，以利于我们今后不断提高教材的出版质量，为广大师生提供更多、更适用的教材。

机械工业出版社

前　言

随着科技的进步和社会的发展，我国对专业人才英语能力的要求越来越高，自动化专业是当今世界发展最迅速、技术更新最活跃的领域之一。我国在该领域注重引进世界先进技术和设备，同时要发展和创造外向型经济，因此该领域对具有专业英语能力人才的需求更加迫切。为了更好地培养学生的专业外语能力，促进具有国际竞争力的人才培养，编者在追求通俗易懂、简明扼要、便于教学和自学的指导思想下对第1版教材进行了改版。

本书由4部分组成，即电力电子技术基础篇、楼宇智能化技术篇、电气自动控制技术篇和工业4.0与中国品牌制造技术篇。本书内容基本涵盖了电力电子技术基础、仪器仪表使用与维护、电能输送、电力系统监控、继电器工作原理、楼宇智能化、办公自动化、可编程逻辑控制器、传感器技术、自动化控制技术、第四次工业革命和具有代表性的中国品牌制造技术等。本书内容丰富、题材广泛，语言通俗易懂，还选入了两篇关于创新名人传记的阅读材料，以激发高职学生学习专业英语的兴趣，并强化他们的创新愿望和能力培养。

本书共分为22个单元，每个单元包括课文、生词、专业术语、长难句解析、翻译技巧和阅读材料。在本书第18单元中增添了以下信息：零件数据库网站介绍与英文网站注册申请表的填写。主要因为随着智能网络和生产系统自动化的深入发展，高职学生的创新设计能力和愿望更加旺盛。而现代的机械、电子和自动化设计早已经脱离了使用图板、圆规等绘图的手动模式，3D设计软件已经普及。很多设计工作是先设计3D图样，然后再转化成2D工程图的形式来完成的。而零件数据库网站有丰富的数据使我们无须再参照纸版说明书中的数据来重新设计已经标准化的零件，甚至零件的3D模型也可以插入到我们的装配图中。这样从零件数据库网站下载的3D数字模型，可以大大提升高职学生的工作效率和设计质量。在本书的最后4个单元的实用英语中，分别用相当篇幅介绍了怎样阅读英文招聘广告、怎样用英文写个人简历、如何写英文求职信等应用文体，以及英语面试过程中的常用技巧，目的是为了使毕业生在就业竞争中能胜人一筹。附录部分汇编了十几篇职业现场的交际对话、各单元的参考译文与部分习题答案（为了培养学生的独立阅读能力，部分阅读材料的参考译文和习题答案将只在电子课件中给出）等内容。

本书可作为高职高专院校电气自动化、生产过程自动化、电力系统自动化和电气控制技术类专业学生的英语教材，也可供相关专业的技术人员使用。每单元参考学时为2~3学时。建议教师根据学生的接受能力和本校学时情况选用本书15~20个单元的内容。附录中的交际英语对话内容，教师可布置给学生主要在课后完成，但教师应有计划地抽查，并占用少量课堂时间做好演示。同时配合生动活泼、灵活多样的互动式教学与课后练习讨论，多方位培养学生的专业英语兴趣与应用能力。对教师在授课中没有选入的单元，学生可根据自己的学习兴趣自学，以拓宽专业英语的知识面。

本书由河南工业职业技术学院徐存善、毛琰虹、党菲菲任主编；周志宇、黄靓、朱理任副主编；蒋志豪、余梦媛、王颖参编。编写分工为：毛琰虹编写第1~4单元，朱理（河南工业职业技术学院）编写第5、6单元，余梦媛（河南工业职业技术学院）编写第7单元，王颖（漯河职业技术学院）编写第8、9单元，黄靓（平顶山工业职业技术学院）编写第10~13单元，党菲菲编写第14~17单元，周志宇（廊坊东方职业技术学院）编写第18~20

单元，徐存善编写第 21~22 单元和附录 3，蒋志豪（河南工业职业技术学院）编写附录 1，附录 2 中各单元参考译文和习题答案分别由对应的编者编写。

本书的编写工作得到了编者所在院校领导的高度重视与大力支持，他们为本书的编写提出了许多宝贵的意见，在此表示衷心的感谢。

由于编者水平有限，加之时间仓促，书中难免有不当之处，恳请广大读者和同行批评指正。

编　者

目　录

出版说明
前言

Chapter Ⅰ　The Base of Power Electronics

Unit 1　Electric Circuits and Electrical Components ······ 1
　Translating Skills：科技英语翻译的标准与方法 ······ 4
　Reading：Electrical Components ······ 6
Unit 2　AC, DC and Electrical Signals ······ 9
　Translating Skills：词义的确定 ······ 12
　Reading for Celebrity Biography（Ⅰ）：Nikola Tesla ······ 13
Unit 3　The Electronic Instruments ······ 16
　Translating Skills：引申译法 ······ 20
　Reading：How to Use a Tester ······ 20
Unit 4　Integrated Circuit ······ 23
　Translating Skills：词性转换 ······ 25
　Reading：Digital Circuit ······ 26
Unit 5　Operational Amplifier ······ 29
　Translating Skills：增词译法 ······ 32
　Reading：Oscillator ······ 33
Unit 6　Heavy and Light Current Engineering ······ 35
　Translating Skills：减词译法 ······ 38
　Reading：Superconductivity ······ 39
Unit 7　How Power Grids Work ······ 42
　Translating Skills：科技英语词汇的结构特征（Ⅰ） ······ 46
　Reading：Transformers ······ 47
Unit 8　Basic Relay Types ······ 50
　Translating Skills：科技英语词汇的结构特征（Ⅱ） ······ 53
　Reading：Fault-Clearing Protective Relays ······ 55
Unit 9　Electric Power System Monitoring ······ 57
　Translating Skills：被动语态的译法 ······ 60
　Reading：Distribution Automation Increases Reliability ······ 61

Chapter Ⅱ　Intelligent Building Technology

Unit 10　Intelligent Building ······ 64
　Translating Skills：非谓语动词 V-ing 的用法 ······ 67
　Reading：Intelligent Hotel ······ 68
Unit 11　Structured Cabling System ······ 71
　Translating Skills：非谓语动词 V-ed 和 to V 的用法 ······ 74

Reading: Solutions of Home Structured Cabling System ... 75
Unit 12　Office Automation System ... 77
Translating Skills: 定语从句的翻译 ... 80
Reading: Multimedia Technology ... 81
Unit 13　Security System ... 84
Translating Skills: 虚拟语气的翻译 ... 87
Reading: Intelligent City ... 88

Chapter Ⅲ　Automatic Control Technology

Unit 14　Introduction to Control Engineering ... 92
Translating Skills: and 引导的句型的译法 ... 95
Reading: Closed-Loop Control System ... 96
Unit 15　Programmable Logic Controller（PLC） ... 97
Translating Skills: 科技英语中一些常用的结构与表达 ... 99
Reading: PLC Programming ... 100
Unit 16　Electronic Measuring Instruments ... 103
Translating Skills: 反译法 ... 106
Reading: Transducers ... 107
Unit 17　Adaptive Control Systems ... 110
Translating Skills: 长难句的翻译 ... 113
Reading: Control System Components ... 115
Unit 18　Automatic Control System ... 116
Useful Information: 零件数据库网站介绍与英文网站注册申请表的填写 ... 120
Reading: Digital Control Systems ... 122
Unit 19　Applications of Automatic Control ... 125
Practical English: 怎样阅读英文招聘广告 ... 128
Reading: VFD system ... 130

Chapter Ⅳ　Industry 4.0 & Chinese Brands Manufacturing

Unit 20　Industry 4.0 Introduction ... 133
Practical English: 怎样用英文写个人简历 ... 137
Reading: Made in China 2025 and Industrie 4.0 Cooperative Opportunities ... 139
Unit 21　China's High-speed Railway ... 143
Practical English: 如何写英文求职信 ... 146
Reading: In Las Vegas, The Chinese Have Arrived ... 147
Unit 22　Sky is the Limit for Drone Manufacturers ... 151
Practical English: 面试技巧 ... 153
Reading for Celebrity Biography（Ⅱ）: Bill Gates ... 156
Appendix ... 159
Appendix 1　English Communication Skills Training for Careers ... 159
Appendix 2　Reference Translations and Keys to Parts of Exercises ... 167
Appendix 3　New Words List ... 206
参考文献 ... 225

Chapter Ⅰ The Base of Power Electronics

Unit 1 Electric Circuits and Electrical Components

Text

Circuits

1. Current

An electric current is a flow of charged particles. Inside a copper wire, current is carried by small negatively-charged particles, called electrons. The electrons drift in random directions until a current starts to flow. When this happens, electrons start to move in the same direction. The size of the current depends on the number of electrons passing per second.

Current is represented by the symbol I, and is measured in amperes, "amps", or A. One ampere is a flow of 6.24×10^{18} electrons per second past any point in a wire. That's more than six million million million electrons passing per second.

In electronic circuits, currents are most often measured in milliamps, mA, that is, thousandths of an amp.

2. Voltage

In the torch circuit, what causes the current to flow? The answer is that the cells provide a "push" which makes the current flow round the circuit. [1]

Each cell provides a push, called its potential difference, or voltage. This is represented by the symbol U, and is measured in volts, or V.

Typically, each cell provides 1.5V. Two cells connected one after another, in series, provide 3V, while three cells would provide 4.5V, as is shown in Fig. 1-1.

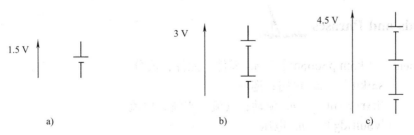

Fig. 1-1 Cells in Series
a) One cell b) Two cells in series c) Three cells in series

3. Cells Connected in Series

Which arrangement would make the lamp glow most brightly? Lamps are designed to work with a particular voltage, but, other things being equal, the bigger the voltage, the brighter the lamp. [2]

Strictly speaking, a battery consists of two or more cells. These can be connected in series, as is usual in a torch circuit, but it is also possible to connect the cells in parallel, like Fig. 1-2.

4. Cells Connected in Parallel

A single cell can provide a little current for a long time, or a big current for a short time. Connecting the cells in series increases the voltage, but does not affect the useful life of the cells. On the other hand, if the cells are connected in parallel, the voltage stays at 1.5V, but the life of the buttery is doubled.

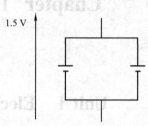

Fig. 1-2　Cells in parallel

A torch lamp which uses 300mA from C-size alkaline cells should operate for more than 20 hours before the cells are exhausted.

5. Resistance

Part of the torch circuit limits, or resists the flow of current. Most of the circuits consist of thick metal conductors which allow current to flow easily. These parts, including the spring, switch plates and lamp connections, have a low resistance. The lamp filament, on the other hand, is made up of very thin wire. It conducts much less easily than the rest of the circuit and has a higher resistance. [3]

The resistance, R, of the filament is measured in ohms, or Ω. If the battery voltage is 3V (2 C-size cells in series) and the lamp current is 300mA, or 0.3A, what is the resistance of the filament?

This is calculated from

$$R = \frac{U}{I} = \frac{3}{0.3} = 10\Omega$$

where R is resistance, U is the voltage across the lamp, and I is current. In this case, 10Ω is the resistance of the lamp filament once it has heated up.

Resistance values in electronic circuits vary from a few ohms, or Ω to values in kilohms, or $k\Omega$ (thousands of ohms) and megohms, or $M\Omega$ (millions of ohms). Electronic components designed to have particular resistance values are called resistors.

New Words and Phrases

　　　　component　[kəm'pəunənt]　*n.* 零件，元件；成分
　　　　circuit　['sə:kit]　*n.* 电路；巡回
　　　　current　['kʌr(ə)nt]　*n.* 流动；气流，水流，电流
　　　　voltage　['vəultidʒ]　*n.* 电压
　　　　resistance　[ri'zistəns]　*n.* 电阻
　　　　electric　[i'lektrik]　*adj.* 导电的，电动的，电的
　　　　charge　[tʃɑ:dʒ]　*v.* 充电，使带电；收费　*n.* 费用，电荷

particle	[ˈpɑːtɪk(ə)l]	n. 粒子；极小量；点
random	[ˈrændəm]	adj. 随机的；任意的；胡乱的 n. 随意
ampere	[ˈæmpeə(r)]	n. 安培
alkaline	[ˈælkəlaɪn]	adj. 碱性的，碱的
filament	[ˈfɪləm(ə)nt]	n. 细丝，灯丝
resist	[rɪˈzɪst]	v. 抵抗，抗拒，反抗
positive	[ˈpɒzɪtɪv]	n. 正数；正面 adj. 积极的；肯定的；真实的；正极的
make up		形成；弥补
in series		串联
in parallel		平行，并联
strictly speaking		严格地说

Notes

[1] The answer is that the cells provide a "push" which makes the current flow round the circuit.

译文：答案是电池单元提供一个"推力"让电流在电路中环行流动。

说明：本句是复合句。句中 that 引导表语从句，但表语从句本身还包含 which makes the current flow round the circuit 这样一个定语从句，修饰先行词 push。

[2] Lamps are designed to work with a particular voltage, but, other things being equal, the bigger the voltage, the brighter the lamp.

译文：灯泡是设计成在特定电压下工作的，但是当其他条件相同时，电压越高，灯泡就越亮。

说明：该句为并列句，两个分句之间用 but 连接。句中 other things being equal 是分词的独立主格结构，在第二个分句中做条件状语。英语中"the more..., the more..."结构用于表示随着前事物的变化，后事物呈相应的变化，译为"越……，越……"。

[3] It conducts much less easily than the rest of the circuit and has a higher resistance.

译文：它（灯泡的灯丝）比电路的其他部分更不容易传导电流，且具有更高的电阻。

说明：句中 it 指代上句中的 the lamp filament。"much less easily than"表示"比……不容易得多"，类似的还有"much less than"表示"远低于"，"much more than"表示"远远超过"。例如：

Recently I eat much less than usual and only with effort.

最近我比往常吃得少，而且要很费劲才能吃下。

Exercises

I. Answer the following questions according to the passage.

1. What is an electric current?
2. What does the size of the current depend on?
3. In the torch circuit, what causes the current to flow?
4. What's the advantage of connecting the cells in series?

Ⅱ. Match the following phrases in column A with column B.

Column A	Column B
1. C-size alkaline cells	a. 随机方向漂移
2. potential difference	b. 并联
3. a flow of charged particles	c. 电气元件
4. connect in series	d. 微小负电荷粒子
5. drift in random directions	e. C 型碱性电池
6. connect in parallel	f. 带电荷的粒子流
7. electrical components	g. 串联
8. small negatively-charged particles	h. 电势差

Ⅲ. Fill in the blanks with the proper word. Change the form if necessary.

active	inductor	interconnection	electric
deliver	generate	voltage	passive

An electric circuit is simply an _____ of the elements. There are two types of elements found in _____ circuits: passive elements and _____ elements. An active element is capable of _____ energy while a _____ element is not. Examples of passive elements are resistors, capacitors, and _____. The most important active elements are _____ or current sources that generally _____ power to the circuit connected to them.

Ⅳ. Translate the following sentences into Chinese.

1. An ideal independent source is an active element that provides a specified voltage or current that is completely independent of other circuit variables.

2. The power absorbed or supplied by a circuit element is the product of the voltage across the element and the current through it.

3. The greater the resistance, the bigger is the voltage needed to send a given current through the wire.

4. Resistors whose resistances do not remain constant for different terminal currents are known as nonlinear resistors.

Translating Skills

科技英语翻译的标准与方法

翻译是再创造，即译者根据原作者的思想，用另一种语言表达出原作者的意思。这就要求译者必须确切理解和掌握原作的内容与含意，在此基础上，很好地运用译文语言把原文的内涵通顺流畅地再现给读者。

1. 翻译的标准

科技英语的翻译标准可概括为"忠实、通顺"4 个字。

忠实，首先指忠实于原文内容，译者必须把原作的内容完整而准确地表达出来，不得任意发挥或增删；忠实还指保持原作风格，尽量表现其本来面目。

通顺，即指译文语言必须通俗易懂，符合规范。

忠实与通顺是相辅相成的，缺一不可。忠实而不通顺，读者会看不懂；通顺而不忠实，脱离原作的内容与风格，通顺就失去了意义。例如：

The electric resistance is measured in Ohms.

误译：电的反抗是用欧姆测量的。

正译：电阻的测量单位是欧姆。

All metals do not conduct electricity equally well.

误译：全部金属不导电得相等好。

正译：并非所有的金属都同样好地导电。

Some special alloy steels should be used for such parts because the alloying elements make them tougher, stronger, or harder than carbon steels.

误译：对这类零件可采用某些特殊的合金钢，因为合金元素能使它们更加坚韧与坚硬。

正译：对这类零件可采用某些特殊的合金钢，因为合金元素能提高钢的韧性、强度、硬度。

从以上例句可以清楚地看到，不能任意删改，并不等于逐词死译；汉语译文规范化，并非是离开原文随意发挥。此外，还应注意通用术语的译法。比如，第 1 句中的 "The electric resistance" 译为 "电阻" 已成为固定译法，不能用别的译法。

2. 翻译的方法

翻译的方法一般来说有直译（literal translation）和意译（free translation）。直译，即指"既忠实于原文内容，又忠实于原文形式"的翻译；意译，就是指忠实于原文的内容，但不拘泥于原文的形式。

翻译应灵活运用上述两种方法，能直译的就直译，需要意译的就意译。对同一个句子来说，有时并非只能用一种方法，可以交替使用或同时并用以上两种方法。

请看下面的例子。

Milky Way, 应译为 "银河"（意译），不可直译为 "牛奶路"。

bull's eye, 应译为 "靶心"（意译），不可直译为 "牛眼睛"。

New uses have been found for old metals, and new alloys have been made to satisfy new demands. 老的金属有了新用途，新的金属被冶炼出来，以满足新的需要（本句前半部分用了意译法，后半部分用了直译法）。

The ability to program these devices will make a student an invaluable asset to the growing electronic industry. 编程这些器件的能力将使学生成为日益增长的电子工业领域中的无价人才（这里 asset 原意为 "资产"，根据上下文意译成 "人才"）。

3. 翻译中的专业性特点

专业英语要求英语与专业内容相互配合，相互一致，这就决定了专业英语与普通英语有很大的差异。专业英语以其独特的语体，明确表达作者在专业方面的见解，其表达方式直截了当，用词简练。即使同一个词，在不同学科的专业英语中含义也是不同的。例如：

The computer took over an immense range of tasks from worker's muscles and brains.

误译：计算机代替了工人大量的肌肉和大脑。

正译：计算机取代了工人大量的体力和脑力劳动。

（这里 muscles and brains 引申为 "体力和脑力劳动"。）

In any cases work doesn't include time, but power does.

误译：在任何情况下，工作不包含时间，但功率包含时间。
正译：在任何情况下，功不包括时间，但功率包括时间。
（这里 work，power 在物理专业分别译为 "功" "功率"。）
Like charges repel each other while opposite charges attracted.
误译：同样的负载相排斥，相反的负载相吸引。
正译：同性电荷相排斥，异性电荷相吸引。
（charge 含义有 "负载、充电、充气、电荷"，按专业知识理解为 "电荷"。）
从以上例句可知，专业英语专业性强，逻辑性强，翻译要力求准确、精练、正式。这不仅要求我们能熟练地运用汉语表达方式，还要求具有较高的专业水平。

Reading

Electrical Components

1. Resistors

A resistor is an electrical component that resists the flow of electrical current. The amount of current (I) flowing in a circuit is directly proportional to the voltage across it and inversely proportional to the resistance of the circuit. This is Ohm's Law and can be expressed as a formula: $I = \dfrac{U_R}{R}$. The resistor is generally a linear device and its characteristics form a straight line when plotted on a graph.

Resistors are used to limit current flowing to a device, thereby preventing it from burning out, as voltage dividers to reduce voltage for other circuits, as transistor biasing circuits, and to serve as circuit loads. [1]

2. Capacitors

Electrical energy can be stored in an electric field. The device capable of doing this is called a capacitor or a condenser.

A simple condenser consists of two metallic plates separated by a dielectric. If a condenser is connected to a battery, the electrons will flow out of the negative terminal of the battery and accumulate on the condenser plate connected to that side. At the same time, the electrons will leave the plate connected to the positive terminal and flow into the battery to make the potential difference just the same as that of the battery. [2] Thus the condenser is said to be charged.

To discharge the condenser, the external circuit of these two plates is completed by joining terminals together with a wire. The electrons start moving from one plate to the other through the wire to restore electrical neutrality.

3. Inductors

An inductor is an electrical device, which can temporarily store electromagnetic energy in the field about it as long as current is flowing through it. [3] Also, inductors are wound with various sizes of wire and in varying numbers of turns which affect the DC (direct current) resistance of the coil.

Excellent information is available about the details of winding coils to desired specifications in The Radio Amateurs Handbook published by the American Radio Relay League (ARRL). Also there are numerous inexpensive special slide rules that allow you to establish required parameters and to read the number of turns, coil length, coil diameter, and so on, needed for the desired results.

New Words and Phrases

  resistor [ri'zistə] n. 电阻器
  electrical [i'lektrikl] adj. 与电有关的，电的
  proportional [prə'pɔʃənl] adj. 比例的，成比例的
  directly proportional 成正比（的）
  inversely proportional 成反比（的）
  capacitor [kə'pæsitə] n. 电容器
  condenser [kən'densə] n. 电容器；冷凝器
  metallic [mi'tælik] adj. 金属的，含金属的
  dielectric [ˌdaii'lektrik] n. 电介质；绝缘体 adj. 非传导性的
  negative terminal 负端
  positive terminal 正端
  discharge [dis'tʃɑ:dʒ] v. 卸货；放电
  neutrality [njuː'træliti] n. 中立，中间状态；中性
  coil [kɔil] n. 卷，圈，线圈 v. 盘绕；把……卷成圈

Notes

[1] Resistors are used to limit current flowing to a device, thereby preventing it from burning out, as voltage dividers to reduce voltage for other circuits, as transistor biasing circuits, and to serve as circuit loads.

译文：电阻器常用作限流器，限制流过器件的电流以防止器件因电流过大而烧坏。电阻器也可以用作分压器，以降低其他电路的电压，如晶体管偏置电路，电阻器还可以用作电路的负载。

说明：这是一个简单句，谓语部分采用了被动语态，科技英语中大量使用被动语态。as 是介词，表示"用于，作为"。

[2] At the same time, the electrons will leave the plate connected to the positive terminal and flow into the battery to make the potential difference just the same as that of the battery.

译文：同时与电源正极相接的极板上的电子将离开极板流入电池正极，这样两极板上就产生了与电池上相等的电势差。

说明：句中 connected to the positive terminal 为过去分词短语作后置定语，修饰 plate。just the same as that of the battery 中的 that 在此处替代的是 potential difference。

[3] An inductor is an electrical device, which can temporarily store electromagnetic energy in the field about it as long as current is flowing through it.

译文：当电流流过电感器时，电感器周围就有电磁场，电感器是以电磁场的形式暂时储存电磁能量的电子器件。

说明：本句为复合句，句中 which 引导了一个非限制性定语从句，对 inductor 进行解释

说明。as long as 表示"只要",引导条件状语从句。

Exercises

I. Decide whether the following statements are True (T) or False (F) according to the text.

1. The amount of current (I) flowing in a circuit is directly proportional to the voltage across it and also directly proportional to the resistance of the circuit.

2. The resistor is generally a linear device and its characteristics form a curve line when plotted on a graph.

3. Resistors are used to limit current flowing to a device.

4. If a condenser is connected to a battery, the electrons will flow out of the negative terminal of the battery and accumulate on the condenser plate connected to the positive terminal.

II. Translate the following sentences into English.

1. 电能可以储存在电场里,能储存电能的元件称为电容器。
2. 一个简单的电容器是由被介质隔开的两块金属板组成的。
3. 电感器由不同尺寸的导线绕制而成,且有不同的匝数,这些都会影响线圈的直流电阻。

Unit 2 AC, DC and Electrical Signals

Text

AC, DC and Electrical Signals

1. Alternating Current (AC)

Alternating Current (AC) flows one way, then the other way, continually reversing direction (as shown in Fig. 2-1 and Fig. 2-2). An AC voltage is continually changing between positive (+) and negative (−). The rate of changing direction is called the frequency of the AC and it is measured in hertz (Hz) which is the number of forwards – backwards cycles per second. [1]

An AC supply is suitable for powering some devices such as lamps and heaters but almost all electronic circuits require a steady DC supply.

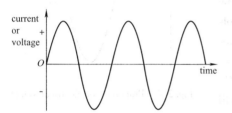   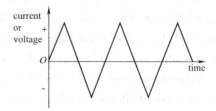

Fig. 2-1 AC from a Power Supply: Fig. 2-2 This triangular signal is AC because it
This shape is called a sine wave.   changes between positive (+) and negative (−).

2. Direct Current (DC)

Direct Current (DC) always flows in the same direction, but it may increase and decrease. A DC voltage is always positive (or always negative), but it may increase and decrease. Electronic circuits normally require a steady DC supply which is constant at one value (as shown in Fig. 2-3). Cells, batteries and regulated power supplies provide steady DC which is ideal for electronic circuits. [2] Lamps, heaters and motors will work with any DC supply.

3. Properties of Electrical Signals

An electrical signal is a voltage or a current which conveys information, usually it means a voltage. The term can be used for any voltage or current in a circuit. The voltage-time graph on the Fig. 2-4 shows various properties of an electrical signal. In addition to the properties labeled on the graph, there is frequency which is the number of cycles per second. [3] The diagram shows a sine wave but these properties apply to any signal with a constant shape.

Amplitude is the maximum voltage reached by the signal. It is measured in volts, V. Peak voltage is another name for amplitude. Peak-to-peak voltage is twice the peak voltage (amplitude). When reading an oscilloscope trace, it is usual to measure peak-to-peak voltage.

Time period is the time taken for the signal to complete one cycle. It is measured in seconds

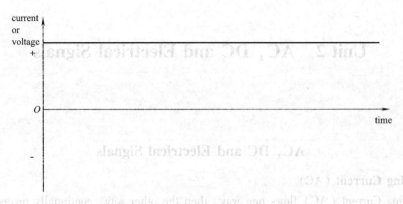

Fig. 2-3 Steady DC: from a battery or regulated power supply, this is ideal for electronic circuits

(s), but time periods tend to be short, so milliseconds (ms) and microseconds (μs) are often used. 1ms = 0.001s and 1μs = 0.000001s.

Frequency is the number of cycles per second. It is measured in hertz (Hz), but frequencies tend to be high, so kilohertz (kHz) and megahertz (MHz) are often used. 1kHz = 1000Hz and 1MHz = 1000000Hz. Frequency = 1/time period and time period = 1/frequency.

Another value used is the effective value of AC. This is the value of alternating voltage or current that will have the same effect on a resistance as a comparable value of direct voltage or current will have on the same resistance.

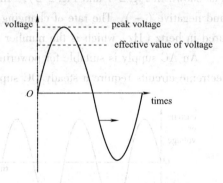

Fig. 2-4 Properties of electrical signals

New Words and Phrases

alternating	[ˈɔːltəneitiŋ]	adj.	交替的；交互的
reverse	[riˈvəːs]	adj. 反面的；颠倒的；反身的	v. 颠倒；倒转
hertz	[həːts]	n.	赫兹（频率单位）
cycle	[ˈsaik(ə)l]	n. 循环；周期	v. 使循环；使轮转
cell	[sel]	n.	细胞；电池；蜂房的巢室；单人小室
regulate	[ˈregjuleit]	vt.	调节，规定；控制；校准
motor	[ˈməutə]	n.	发动机，马达；汽车
label	[ˈleib(ə)l]	n. 标签；商标；签条	v. 标注；贴标签于
sine	[sain]	n.	正弦
amplitude	[ˈæmplitjuːd]	n.	振幅；丰富，充足
oscilloscope	[əˈsiləskəup]	n.	示波器
trace	[treis]	v. 追踪，查探；追溯	n. 痕迹，踪迹
millisecond	[ˈmilisek(ə)nd]	n.	毫秒；千分之一秒
microsecond	[ˈmaikrəuˌsekənd]	n.	微秒；一百万分之一秒
megahertz	[ˈmegəhəːts]	n.	兆赫兹

Technical Terms

Alternating Current　交流电
Direct Current　直流电
power supply　电源
peak voltage　峰值电压
peak-to-peak voltage　峰-峰值电压
effective value　有效值

Notes

［1］The rate of changing direction is called the frequency of the AC and it is measured in hertz (Hz) which is the number of forwards-backwards cycles per second.

译文：这种变换方向的速率称为交流电的频率，测量单位是赫兹，它表示一秒内（交流电）正反向周期性变化的次数。

说明：本句是复合句。句中 which is the number of forwards-backwards cycles per second 为定语从句，修饰先行词 hertz。短语 cycles per second 意为"每秒循环数"。

［2］Cells, batteries and regulated power supplies provide steady DC which is ideal for electronic circuits.

译文：干电池、蓄电池和稳压电源能提供对于电子电路来说理想的、稳定的直流电。

说明：此句是复合句。句中 which is ideal for electronic circuits 为定语从句，修饰先行词 DC。

［3］In addition to the properties labeled on the graph, there is frequency which is the number of cycles per second.

译文：除了图上标示的特性外，还有频率，它表示（电信号）每秒钟的周期数。

说明：该句是复合句。句中 labeled on the graph 为过去分词短语做后置定语，修饰 properties。which 引导定语从句，修饰先行词 frequency。

Exercises

Ⅰ. **Answer the following questions according to the passage.**

1. What is an AC supply suitable for?
2. What is an electrical signal?
3. What's the relation of peak-to-peak voltage and amplitude?
4. When can we measure peak-to-peak voltage?

Ⅱ. **Fill in the blanks.**

1. An AC voltage is continually changing between ＿＿＿＿＿＿ and ＿＿＿＿＿＿.
2. Direct Current (DC) always flows in the same direction, but it may ＿＿＿＿＿＿ and ＿＿＿＿＿＿.
3. Amplitude is the ＿＿＿＿＿＿ voltage reached by the signal. It is measured in ＿＿＿＿＿＿, V. ＿＿＿＿＿＿ is another name for amplitude.

11

4. Time period is the time taken for the _____ to complete one cycle, It is measured in seconds (s), but time periods tend to be short, so _____ and _____ are often used.

Ⅲ. Multiple choice.

1. An electric current is the flow of _____.
   A. sort   B. stream
   C. energy   D. electrons

2. The electrons first flow in one direction and then in the other in _____ current.
   A. direct   B. alternating
   C. flowing   D. sine wave

3. The number of cycles per second is the wave's _____ which is measured in hertz.
   A. frequency   B. period
   C. IC   D. AC

4. Which of the following material is not conductor?
   A. copper   B. glass
   C. water   D. iron

Ⅳ. Translate the following short passage into Chinese.

The current that flows steadily in one direction is usually called a direct current. We know that the electrical system in an automobile and an air plane, the telegraph, the telephone and the trolly-bus use the direct current. Direct current is also used to meet some of the industrial requirements. For industry and many other purposes, however, mostly, cities make use of another type of electric current which flows first in one direction and then in another. It was given the name of an alternating current. We know that the alternating current is the very current that makes radio possible.

Translating Skills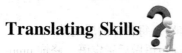

词义的确定

英、汉两种语言都有一词多性、一词多义的现象。一词多性是指一个词往往具有多个词性，具有几种不同的意义。例如，display 既可作名词表示"显示（器）"，又可以作形容词表示"展览的、陈列用的"，还可作动词表示"显示、表现"等意思。一词多义是说在同一词性中，往往有几个不同的词义。例如，power 这个词，作名词的意思包括"电力、功率、次方"等。在翻译过程中，在弄清句子结构后，就要善于选择和确定句子中关键词的词性和意思。选择和确定词义通常从以下几个方面入手。

1. 根据词性确定词义

确定某个词的词义时，首先要确定这个词在句中应属于哪一种词性，然后再进一步确定其词义。下面以 display 为例：

Here, you have the option of defining your own display variants. 这里，你有权定义你自己的显示形式。（display 为名词）

Often, it is best to display materials on an information table. 通常，最好把资料放在提供各类信息的桌子上展示。（display 为动词）

12

The reverse side of a control panel, display panel, or the like is the side with the interconnecting wiring. 控制面板、显示面板或类似的面板的反面，是带有互连接线的那一面。(display 为形容词)

2. 同一词性表达不同词义

英语中的词，即使属于同一词性，在不同的场合中也往往具有不同的含义。此时，必须根据上、下文的联系及整个句子的意思加以判断和翻译。例如 as 这个词作连词时有以下用法：

The volume varies as the temperature increases. 体积随着温度增加而变化。(as 引导时间状语从句)

As heat makes things move, it is a form of energy. 因为热能使物体运动，所以热是能的一种形式。(as 引导原因状语从句)

3. 根据单词搭配情况确定词义

英译汉时，不仅必须根据上、下文的联系理解词义，还需要根据词的搭配情况来理解词义。尤其在科技文献中，由于学科及专业不同，同一个词在不同的专业中具有不同的意义。比如，

The fifth power of two is thirty-two. 2 的 5 次方是 32。(数学)

With the development of electrical engineering, power can be transmitted over long distances. 随着电气工程的发展，电力能输送得非常远。(电学)

Friction can cause a loss of power in every machine. 摩擦能引起每一台机器的功率损耗。(物理学)

4. 根据名词的单复数选择词义

英语中有些名词的单数或复数表达的词义完全不同。例如：

名　　词	单 数 词 义	复 数 词 义
facility	简易，灵巧	设施，工具
charge	负荷，电荷	费用
spirit	精神	酒精

Although they lost, the team played with tremendous spirit. 他们虽然输了，但却表现出了极其顽强的精神。

Whisky, brandy, gin and rum are all spirits. 威士忌、白兰地、杜松子酒和朗姆酒都是烈酒。

Reading for Celebrity Biography（Ⅰ）

Nikola Tesla

Nikola Tesla was born in 1856 in Smiljan Lika, Croatia. He was the son of a Serbian Orthodox clergyman. Tesla studied engineering at the Austrian Polytechnic School. He worked as an electrical engineer in Budapest and later emigrated to the United States in 1884 to work at the Edison Machine Works. He died in New York City on January 7, 1943.

During his lifetime, Tesla invented fluorescent lighting, the Tesla induction motor, the Tesla coil, and developed the alternating current (AC) electrical supply system that included a motor and transformer, and 3-phase electricity.

Tesla is now credited with inventing modern radio as well; since the Supreme Court overturned Guglielmo Marconi's patent in 1943 in favor of Nikola Tesla's earlier patents. When an engineer (Otis Pond) once said to Tesla, "Looks as if Marconi got the jump on you" regarding Marconi's radio system, Tesla replied, "Marconi is a good fellow. Let him continue. He is using seventeen of my patents."

The Tesla coil, invented in 1891, is still used in radio and television sets and other electronic equipment.

Nikola Tesla —Mystery Invention

Ten years after patenting a successful method for producing alternating current, Nikola Tesla claimed the invention of an electrical generator that would not consume any fuel. [1] This invention has been lost to the public. Tesla stated about his invention that he had harnessed the cosmic rays and caused them to operate a motive device.

In total, Nikola Telsa was granted more than one hundred patents and invented countless unpatented inventions.

Nikola Tesla and George Westinghouse

In 1885, George Westinghouse, head of the Westinghouse Electric Company, bought the patent rights to Tesla's system of dynamos, transformers and motors. Westinghouse used Tesla's alternating current system to light the World's Columbian Exposition of 1893 in Chicago.

Nikola Tesla and Thomas Edison

Nikola Tesla was Thomas Edison's rival at the end of the 19th century. In fact, he was more famous than Edison throughout the 1890's. His invention of polyphase electric power earned him worldwide fame and fortune. At his zenith he was an intimate of poets and scientists, industrialists and financiers. Yet Tesla died destitute, having lost both his fortune and scientific reputation. [2] During his fall from notoriety to obscurity, Tesla created a legacy of genuine invention and prophecy that still fascinates today.

New Words and Expressions

polytechnic　　[ˌpɒlɪˈteknɪk]　　adj. 综合技术的 n. 理工专科学校
emigrate　　['emɪɡreɪt]　　vi. 移居；移居外国 vt. 移民
fluorescent　　[fluəˈres(ə)nt]　　adj. 荧光的；发亮的 n. 荧光；荧光灯
overturn　　[ˌəʊvəˈtɜːn]　　vt. 推翻；倾覆；破坏
motive　　['məʊtɪv]　　n. 动机，目的；adj. 发动的；成为动机的
grant　　[ɡrɑːnt]　　vt. 授予；允许；承认 vi. 同意
dynamo　　['daɪnəməʊ]　　n. 发电机；精力充沛的人
transformer　　[trænsˈfɔːmə, trɑːns-; -nz-]　　n. [电] 变压器
polyphase　　['pɒlɪfeɪz]　　adj. 多相的
zenith　　['zenɪθ]　　n. 顶峰；顶点；最高点
intimate　　['ɪntɪmət]　　adj. 亲密的；私人的；精通的；n. 知己；至交

destitute	[ˈdestɪtjuːt]	adj. 穷困的；缺乏的 n. 赤贫者 vt. 使穷困；夺去
notoriety	[nəʊtəˈraɪɪtɪ]	n. 声名狼藉；名声远扬；著名人物
obscurity	[əbˈskjʊərɪtɪ]	n. 朦胧；阴暗；晦涩；身份低微；不分明
prophecy	[ˈprɒfɪsɪ]	n. 预言；预言书；预言能力
coupling	[ˈkʌplɪŋ]	n. [电] 耦合；结合，联结 v. 连接
legacy	[ˈlegəsɪ]	n. 遗赠，遗产

Technical Terms

alternating current　多相电
electrical generator　发电机
cosmic ray　宇宙射线

Notes

[1] Ten years after patenting a successful method for producing alternating current, Nikola Tesla claimed the invention of an electrical generator that would not consume any fuel.

译文：在发明了产生交流电的方法并申请专利10年后，尼古拉·特斯拉声称这是发明了一种不需要任何燃料的发电机。

说明：这是个包含有定语从句的复合句。主句是 Nikola Tesla claimed … generator，that 引导定语从句修饰 generator。Ten years…current 是时间状语，修饰 claimed。其中 for producing… 是介词短语作后置定语修饰 method，after patenting… method 也是介词短语作后置定语修饰 Ten years。

[2] Yet Tesla died destitute, having lost both his fortune and scientific reputation.

译文：然而特斯拉去世时贫困潦倒，丢失了财富和在科学界的声誉。

说明：这是一个简单句。Died 是不及物动词，destitute 是形容词作方式状语，句中 having lost… 表示伴随情况的结果状语。

Exercises

I. Decide whether the following statements are True (T) or False (F) according to the text.

1. Nikola Tesla is the inventor of polyphase alternating current.
2. Tesla was very rich when he died at the year of 1943.
3. Tesla once worked in Edison Machine Works and he was Thomas Edison's rival at the end of the 19th century.
4. Tesla made friends with many successful man at his zenith.

II. Translate the following sentences into English.

1. 特斯拉声称他每天从凌晨3点工作到晚上11点，星期天和节假日也不例外。
2. 1890年后特斯拉试验用电感和电容与高压交流电耦合的方式传输他的特斯拉线圈生成的电力。
3. 尝试着发明一个生成交流电的更好的方法，特斯拉发明了一个蒸汽驱动的往复式电力发电机，并在1893年申请专利，同年在哥伦布纪念博览会上开始使用。

Unit 3　The Electronic Instruments

Text

Multimeters and Oscilloscopes

1. Multimeters

A multimeter is a general-purpose meter capable of measuring DC and AC voltage, current, resistance, and in some cases, decibels. There are two types of meters: analog, using a standard meter movement with a needle (Seen in Fig. 3-1a), and digital, with an electronic numerical display (Seen in Fig. 3-1b). Both types of meters have a positive (+) jack and a common jack (-) for the test leads, a function switch to select DC voltage, AC voltage, DC current, AC current, or ohms and a range switch for accurate readings. The meters may also have other jacks to measure extended ranges of voltage (1 to 5 kV) and current (up to 10A). There are some variations to the functions used for specific meters.

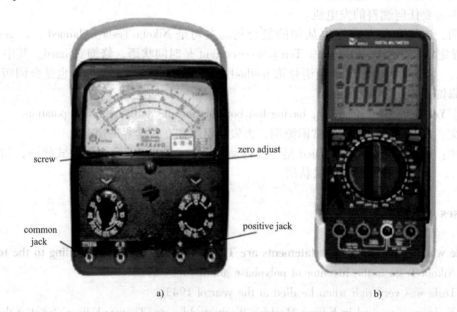

Fig. 3-1　Portable multimeters
a) Analog multimeter　b) Digital multimeter

Besides the function and range switches (sometimes they are in a single switch), the analog meter may have a polarity switch to facilitate reversing the test leads. The needle usually has a screw for mechanical adjust to set it to zero and also a zero adjust control to compensate for weakening batteries when measuring resistance[1]. An analog meter can read positive and negative voltage by simply reversing the test leads or moving the polarity switch. A digital meter usually has an automatic indicator for polarity on its display.

Meters must be properly connected to a circuit to ensure a correct reading. A voltmeter is always placed across (in parallel) the circuit or component to be measured. When measuring current, the circuit must be opened and the meter inserted in series with the circuit or component to be measured. When measuring the resistance of a component in a circuit, the voltage to the circuit must be removed and the meter placed in parallel with the component.

2. The Oscilloscope

The oscilloscope (Seen in Fig. 3-2) is basically a graph-displaying device—it draws a graph of an electrical signal. When the signal is inputted into the oscilloscope, an electron beam is created, focused, accelerated, and properly deflected to display the voltage waveforms on the face of a cathode-ray tube (CRT)[2].

In most applications the graph shows how signals change over time: the vertical (Y) axis represents voltage and the horizontal (X) axis represents time. The amplitude of a voltage waveform on an oscilloscope screen can be determined by counting the number of centimeters (cm), vertically, from one peak to the other peak of the waveform (Seen in Fig. 3-3) and the multiplying it by the setting of the V/cm control[3]. As an example, if the amplitude was 5cm and the control was set on 1V/cm, the peak-to-peak voltage would be 5V.

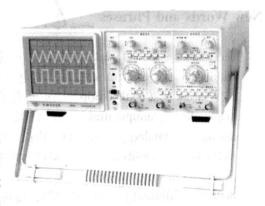

Fig. 3-2 Dual-trace oscilloscope

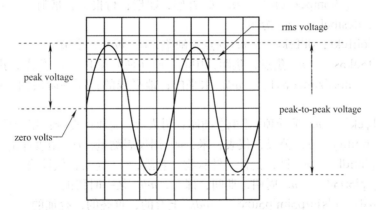

Fig. 3-3 Peak voltage and peak-to-peak voltage

Time can be measured using the horizontal scale of the oscilloscope. Time measurements include measuring the period, pulse width and frequency. Frequency is the reciprocal of the period, so once you know the period, the frequency is divided by the period.

The frequency of a waveform can be determined by counting the number of centimeters, horizontally, in one cycle of the waveform and the multiplying it by the setting time/cm control. For example, if the waveform is 4cm long and the control is set at 1ms/cm, the period would be 4ms. The frequency can now be found from the formula:

$$f = \frac{1}{P} = \frac{1}{4\text{ms}} = 250\text{Hz}$$

If the control was set at 100μs/cm, the period would be 400μs and the frequency would be 2.5kHz.

A dual-trace oscilloscope is advantageous to show the input signal and output signal of one circuit in the same time, to determine any defects, and indicate phase relationships. The two traces may be placed over each other (superimposed) to indicate better the phase shift between two signals.

New Words and Phrases

accelerate   [əkˈseləreit]   vt. 使……加快, vi. 加速, 促进, 增加
multimeter   [ˈmʌltimiːtə]   n. [电] 万用表, 数字万用表, 多用电表
oscilloscope [əˈsiləskəup]   n. [电子] 示波器; 示波镜
amplitude   [ˈæmplitjuːd]   n. 广阔, 丰富, 振幅
analog   [ˈænəlɒg]   n. [自] 模拟; 类似物 adj. [自] 模拟的; 有长短针的
facilitate   [fəˈsiliteit]   vt. 促进, 帮助, 使容易
insert   [inˈsəːt, ˈinsəːt] vt. 插入, 嵌入, n. 插入物
deflect   [diˈflekt]   vt. 使偏斜, 使转向, 使弯曲, vi. 偏斜, 转向
multiply   [ˈmʌltiplai]   vt. 乘; 使增加 vi. 乘; 增加 adv. 多样地 adj. 多层的; 多样的
reciprocal   [riˈsiprəkəl]   adj. 相互的, 倒数的, n. 倒数
beam   [ˈbiːm]   n. 梁, (光线的) 束, 电波, v. 播送
compensate   [ˈkɒmpenseit]   vt. vi. 补偿, 赔偿, 付报酬; 抵消
decibel   [ˈdesibel]   n. 分贝
defect   [ˈdiːfekt, diˈfekt] n. 缺点, 缺陷, vi. 叛变, 变节
focus   [ˈfəukəs]   n. 焦点, 焦距, 中心, v. (使) 聚焦, n. 焦点, 焦距
horizontal   [hɒriˈzɒnt(ə)l] adj. 水平的; 地平线的; 同一阶层的 n. 水平线, 水平面
jack   [dʒæk]   n. 千斤顶; [电] 插座; 男人 vt. 增加; 提醒; 用千斤顶顶起某物
meter   [ˈmiːtə]   n. 公尺, 仪表, 米, vt. 用仪表测量, vi. 用表计量
needle   [ˈniːdl]   n. 针, 指针, 针状物, 刺激, vi. 缝纫, 做针线
period   [ˈpiəriəd]   n. 周期, 期间, 课时, adj. 某一时代的
superimposed [ˈsjuːpəimˈpəuzd] adj. 上叠的, 重叠的, 叠加的
vertical   [ˈvəːtikəl]   adj. 垂直的, 直立的, 头顶的, 顶点的, n. 垂直线, 垂直面

Technical Terms

cathode-ray tube (CRT)   阴极射线显像管
dual-trace oscilloscope   双踪示波器
peak-to-peak voltage   电压峰－峰值
phase shift   相位漂移/差别, 移相
polarity switch   极性开关
rms voltage   电压有效值

Notes

[1] The needle usually has a screw for mechanical adjust to set it to zero and also a zero adjust control to compensate for weakening batteries when measuring resistance.

译文：指针常常有一个旋钮来机械调零。当测量电阻时，一个零点调节控制（钮）用来对电池电压的不足作出补偿调节（即保证电阻为 0 时指针指向零值）。

说明：句中 to compensate for weakening batteries 作后置定语。

[2] When the signal is inputted into the oscilloscope, an electron beam is created, focused, accelerated, and properly deflected to display the voltage waveforms on the face of a cathode-ray tube (CRT).

译文：当信号输入到示波器中时，就产生一个电子束，该电子束被聚焦、加速并适当偏离，在阴极射线管的显示屏上显示电压的波形。

说明：句中 to display the voltage waveforms…作结果状语。

[3] The amplitude of a voltage waveform on an oscilloscope screen can be determined by counting the number of centimeters (cm), vertically, from one peak to the other peak of the waveform (Seen in Fig. 3-3) and the multiplying it by the setting of the V/cm control.

译文：示波器屏幕上电压波形的幅度可以通过数出电压波峰与波谷之间纵向距离的厘米数来确定（图 3-3），将这个厘米数乘以 V/cm 控制钮的设定值就得到电压的幅度值。

说明：句中 from one peak…作后置定语，修饰 centimeters。by counting…and the multplying it by…是方式状语，修饰 be determined。

Exercises

I. Answer the following questions according to the text.

1. What can be measured with a multimeter?
2. What is the function of an oscilloscope?
3. Do you know the differences between a digital multimeter and an analog multimeter? Can you describe them?
4. What is the advantage of a dual-trace oscilloscope?

II. Translate the following phrases into Chinese.

1. general-purpose meter   2. reverse the test leads
3. mechanical adjust   4. voltage amplitude
5. dual-trace oscilloscope   6. signal generator
7. analog multimeter   8. phrase relationship
9. display the voltage waveform 10. positive voltage

III. Translate the following sentences into English.

1. 模拟万用表有一个极性开关，可以很方便地交换测试笔的极性。
2. 对特殊的万用表而言，还有一些其他功能的变化。
3. 示波器显示一个电子信号的图像。
4. 双踪示波器具有同时显示输入信号和输出信号的优点。

5. 两路信号的波形重叠在一起能较好地显示输入与输出信号相位的差别。

Translating Skills

引 申 译 法

英、汉两种语言在表达上有很大差别。翻译时，有些词或词组不能直接搬用词典中的释义，若生搬硬套，会使译文生硬晦涩，难以看懂，甚至造成误解。所以，要在弄清原文内涵的基础上，根据上下文的逻辑关系和汉语的搭配习惯，对词义加以引申。若遇到专业方面的内容，必须选用专业术语。引申后的词义能更确切地表达原文意义。例如：

However, colors can give more force to the form of the product.
欠佳译法：然而，色彩能给予产品更多的力量。
引申译法：然而，色彩能使产品外形增添美感。
Power plugs are male electrical connectors that fit into female electrical sockets.
欠佳译法：电源插头是雄性连接器适配雌性连接器。
引申译法：电源插头可以插入电源插座。
High-speed grinding does not know this disadvantage.
欠佳译法：高速磨床不知道这个缺点。
引申译法：高速磨床不存在这个缺点。
The charge current depends upon the technology and capacity of the battery being charged.
欠佳译法：充电电流的大小取决于充电技术和被充电电池的容量。
引申译法：充电电流的大小根据充电技术和电池容量的不同而不同。

Reading

How to Use a Tester

This instrument is designed to use for measuring DC voltage, measuring AC voltage, measuring resistance, conductivity test and diode test. It has $3\left(\frac{1}{2}\right)$ digits liquid crystal display. So it is called a digital multi-meter.

The following is illustrating its operation.

1. Measuring DC voltage（Seen in Fig. 3-4）

1）Set the function switch to "DCV".
2）Connect the test leads to the circuit to be measured.
3）Read the display.

Note：
1）" - "（minus sign）is displayed when the polarity of the test leads is reversed.
2）Use the test leads with the normal polarity when measuring a voltage that includes spike pulses（such as horizontal output signal of a TV set）.

2. **Measuring AC Voltage** (Seen in Fig. 3-5)
    1) Set the function switch to "ACV".
    2) Connect the test leads to the circuit to be measured.
    3) Read the display.

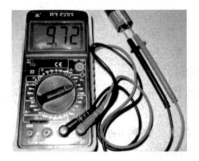 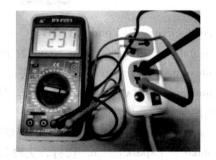

Fig. 3-4 Measuring DC voltage   Fig. 3-5 Measuring AC voltage

Note: It is not necessary to consider the polarity of the test leads.

3. **Measuring Resistance** (Seen in Fig. 3-6)
    1) Set the function switch to "Ω".
    2) Connect the test leads to the circuit to be measured.
    3) Read the display.

Note: Be sure to turn off the power of the circuit to be measured before connecting the leads.

Fig. 3-6 Measuring resistance

4. **Conductivity Test** (Seen in Fig. 3-7)
    1) Set the function switch to "·))".
    2) Connect the test leads to the circuit to be tested.
    3) Conductivity is good when the buzzer beeps and the mark "·))" is displayed.

5. **Diode Test** (Seen in Fig. 3-8)

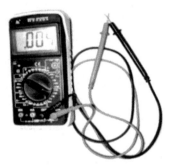 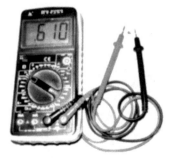

Fig. 3-7 Conductivity test   Fig. 3-8 Diode test

1) Set the function switch to "·))".
2) With a normal diode, the display shows the forward resistance of the diode when the black

test lead is connected to the cathode of the diode and the red test lead to the anode; it displays "1." when the test leads are reversed.

3) When the test leads are open, the display reads "1.".

New Words and Phrases

  conductivity [kɒndʌk'tɪvɪtɪ]   *n.* 导电性［物］［生理］传导性，电阻率
  digital ['dɪdʒɪt(ə)l]   *adj.* 数字的；手指的 *n.* 数字；键
  illustrate ['iləstreit]   *vt.* 阐明，举例说明，图解
  function ['fʌŋ(k)ʃ(ə)n]   *n.* 功能；［数］函数 *vi.* 运行；行使职责 *vi.* 运行，起作用
  display [dɪ'spleɪ]   *n.* 显示；炫耀 *vt.* 显示；表现；陈列 *adj.* 展览的
  spike [spaik]   *n.* 长钉，尖峰信号，*vt.* 用尖物刺穿
  buzzer ['bʌzə]   *n.* 蜂鸣器；嗡嗡作声的东西；信号手
  beep [bi:p]   *vi.* 嘟嘟响，*n.* 哔哔的声音，警笛声
  cathode ['kæθəud]   *n.* 阴极，负极
  multi-meter   万用电表，多量程仪表
  DC (Direct Current)   直流
  test leads   表笔

Exercises

Ⅰ. Translate the following sentences into English.
 1. 该仪表是用于测量电压、电阻、导电性和二极管的。
 2. 该仪表的使用方法示例如下。
 3. 表笔极性接反时，会出现"-"（负号）。
 4. 把测试表笔接在被测电路上。

Ⅱ. Practice how to use a tester.

Unit 4 Integrated Circuit

Text

An integrated circuit (IC) is a combination of a few interconnected circuit elements such as transistors, diodes, capacitors and resistors. It is a small electronic device made out of a semiconductor material. The first integrated circuit was developed in the 1950s by Jack Kilby of Texas Instruments and Robert Noyce of Fairchild Semiconductor.

The electrically interconnected components that make up an IC are called integrated elements.[1] If an integrated circuit includes only one type of components, it is said to be an assembly or set of components.

Integrated circuits (Seen in Fig. 4-1) are used for a variety of devices, including microprocessors, audio and video equipments, and automobiles. Integrated circuits are often classified by the number of transistors and other electronic components, they contain:

· SSI (small-scale integration): Up to 100 electronic components per chip;

Fig. 4-1 Integrated circuits

· MSI (medium-scale integration): From 100 to 3000 electronic components per chip;
· LSI (large-scale integration): From 3000 to 100,000 electronic components per chip;
· VLSI (very large-scale integration): From 100,000 to 1,000,000 electronic components per chip;
· ULSI (ultra large-scale integration): More than 1 million electronic components per chip.

As the capability to integrate a greater number of transistors in a single integrated circuit (IC) grows, it is becoming more common that an application-specific IC (ASIC) is required, at least for high volume applications.[2] Advances in silicon technology have allowed IC designers to integrate more than a few million transistors on a chip; even a whole system of moderate complexity can now be implemented on a single chip.

The invention of IC is a great revolution in the electronic industry. Sharp size, weight reductions are possible with these techniques, and more importantly, high reliability, excellent functional performance, low cost and low power dissipation can be achieved. ICs are widely used in the electronic industry.

New Words and Expressions

assembly　　[ə'semblɪ]　　n. 装配；集会，集合
combination　　[ˌkɒmbɪ'neɪʃ(ə)n]　　n. 结合，联合，合并
interconnect　　[ˌɪntəkə'nekt]　　vt. 使互相连接　vi. 互相联系

semiconductor	[ˌsemɪkən'dʌktə]	n.	[电子][物]半导体；半导体器件
microprocessor	[maɪkrə(ʊ)'prəʊsesə]	n.	[计]微处理器
equipment	[ɪ'kwɪpm(ə)nt]	n.	设备，装备；器材，配件
automobile	['ɔːtəməbiːl]	n.	<美>汽车 vt. 驾驶汽车
contain	[kən'teɪn]	vt.	包含，容纳；克制，遏制；牵制
classify	['klæsɪfaɪ]	vt.	分类，归类；把……列为密件
video	['vɪdɪəʊ]	n.	[电子]视频；录像，录像机 adj. 视频的；录像的
implement	['ɪmplɪm(ə)nt]	vt.	实施，执行；实现，使生效 n. 工具；手段
ultra	['ʌltrə]	adj.	极端的；过分的 n. 极端主义者；激进论者
silicon	['sɪlɪk(ə)n]	n.	[化学]硅；硅元素
moderate	['mɒd(ə)rət]	adj.	有节制的；温和的；适度的；中等的
complexity	[kəm'pleksətɪ]	n.	复杂性，复杂的事物；复合物
revolution	[revə'luːʃ(ə)n]	n.	革命；彻底改变；旋转；运行，公转
industry	['ɪndəstrɪ]	n.	工业；产业（经济词汇）；工业界
reliability	[rɪlaɪə'bɪlɪtɪ]	n.	信度；可靠性，可靠度
dissipation	[dɪsɪ'peɪʃ(ə)n]	n.	（物质、精力逐渐的）消散，分散功耗
functional	['fʌŋ(k)ʃən(ə)l]	adj.	功能的

the circuit element　电路元件
the video equipment　视频设备
ultra large-scale integration（ULSI）　超大规模集成电路
the electronic industry　电子工业

Notes

[1] The electrically interconnected components that make up an IC are called integrated elements.

译文：电学上把组成集成电路的彼此相连的元器件称为集成元器件。

说明：句中that引导定语从句，that make up an IC为定语从句，修饰先行词interconnected components。

[2] As the capability to integrate a greater number of transistors in a single integrated circuit (IC) grows, it is becoming more common that an application-specific IC (ASIC) is required, at least for high volume applications.

译文：随着在一个芯片上集成大量晶体管的能力（即集成电路的集成量）的提高，对专用集成电路的需求已更加普遍。至少对大批量的应用来说更需要专用的集成电路。

说明：这是一个复合句，as引导一个时间状语从句；主句本身也是一个复合句，it是形式主语，真正的主语是由that引导的主语从句来担任的；at least引导让步状语。

Exercises

Ⅰ. Answer the following questions.

1. What is an integrated circuit? When was the first integrated circuit developed?

2. What are called integrated elements? What is named an assembly or set of components?
3. How many types are integrated circuits often classified? What are they?
4. What are possibly achieved with the invention of IC?

II. Decide whether the following statements are True (T) or False (F) according to the text.
1. An integrated circuit is a small electronic device made out of a semiconductor material.
2. A whole system of moderate complexity can now be implemented on a single chip.
3. Integrated circuits are used for a few of devices.
4. Integrated circuits are often classified by the number of transistors and other electronic components.
5. VLSI is up to from 3000 to 10000 electronic components per chip.
6. The invention of IC is not a great revolution in the electronic industry.

III. Translate the following sentences into English.
1. 集成电路是几种相互连接的电路元器件在一块半导体芯片上的组合。
2. 把组成集成电路的彼此电气连接的元器件称为集成元器件。
3. 集成电路已广泛应用在电子工业。
4. 集成电路通常根据其包含的晶体管和其他电路元器件的数量来归类。

IV. Translate the following short passage into Chinese.
An integrated circuit looks like nothing more than a tiny chip of metal, perhaps one-half of a centimeter on a side, and not much thicker than a sheet of paper. It is so small that if it fell on the floor, it could be easily swept up with the dust. Although it is very small, it represents the most highly skilled technology at every step of its manufacture. At today's level of development, it might comprise more than ten thousand even several millions of separate electronic elements including elements of many different functions, such as diodes, transistors, capacitors and resistors.

Translating Skills

词 性 转 换

在翻译过程中，由于英、汉两种语言的表达方式不同，不能逐词对译。原文中有些词在译文中须转换词性，才能使译文通顺流畅。词性转换包括以下几种情况。

1. 英语动词、形容词、副词译成汉语名词

Telecommunications <u>means</u> so much in modern life that without it our modern life would be impossible. 电信在现代生活中<u>意义</u>重大，没有它就不可能有我们现在的生活。

The cutting tools must be <u>strong, tough, hard</u> and <u>wear resistant</u>. 刀具必须有足够的<u>强度，韧性，硬度</u>，而且耐磨。

Dynamics is divided into statics and kinetics, <u>the former treating of forces in equilibrium, the latter</u> of the relation of force to motion. 力学分为静力学和动力学：<u>前者研究平衡力，后者研究力和运动的关系</u>。

The image must be <u>dimensionally</u> correct. 图形的<u>尺寸</u>必须正确。

2. 英语名词、介词、形容词、副词译成汉语动词

<u>Substitution</u> of manual finishing is one example of HSM application. <u>替代</u>手工精加工是高速

加工应用的一个例子。

Scientists are confident that all matter is indestructible. 科学家们深信一切物质是不灭的。

In any machine input work equals output work plus work done against friction. 任何机器的输入功，都等于输出功加上克服摩擦所做的功。

Open the valve to let air in. 打开阀门，让空气进入。

3. 英语的名词、副词和动词译成汉语形容词

This wave guide tube is chiefly characterized by its simplicity of structure. 这种波导管的主要特点是结构简单。

They said that such knowledge is needed before they can develop a successful early warning system for earthquakes. 他们说，这种知识对他们发明一种有效的地震早期警报是必要的。

4. 英语形容词、名词译成汉语副词

With slight repairs the television transmitters can be used. 只要稍加修理，这台视频发射机就可使用。

A continuous increase in the temperature of a gas confined in a container will lead to a continuous increase in the internal pressure within the gas. 不断提高密封容器内气体的温度，会使气体的内压力不断增大。

Reading

Digital Circuit

The phrase "digital electronics" is used to describe those circuit systems which primarily operate with the use of only two different voltage levels or two other binary states. [1] The two different states by which digital circuits operate may be of several forms. They can, in the most simple form, consist of the opening and closing of a switch. In this case, the closed-switch state can be represented by 1 and the open-switch state by 0. [2]

A very common method of digital operation is achieved by using voltage pulses. The presence of a positive pulse can be represented by 1 and the absence of a pulse by 0. With a square-wave signal, the positive pulses can represent 1 and the negative pulses can represent 0.

Integrated circuits containing many transistors are most commonly used as switching devices in digital electronics logic gates. The three basic types of digital logic gates are the AND gate, the OR gate, and the NOT gate (as shown in Fig. 4-2). The operation of an AND gate is mathematically expressed by the equation $A \cdot B = C$. This can be read as "input A and input B equals output C". The operation of an OR gate is often expressed by the equation $A + B = C$. This can be read as "either input A or input B (or both) equals output C". An important function of NOT gate is to produce signal inversion or an output signal that is opposite in nature to the input signal. [3] Any logic function can be performed by the three basic gates that have been described. Even in a large scale digital system, such as a computer, control or digital-communication system, there are only a few basic operations, which must be performed. [4] The three basic types of digital logic gates and the flip-flop are the four circuits most commonly employed in such systems.

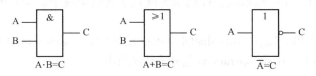

Fig. 4-2　Digital logic Gates

New Words and Phrases

primarily　　['praɪm(ə)rɪlɪ]　　adv. 主要地，根本上；首先；最初的
binary　　['baɪnərɪ]　　adj. 二元的，二态的；二进制的
pulse　　[pʌls]　　n. 脉搏；脉冲　vt. 使跳动　vi. 跳动，脉跳
represent　　[ˌreprɪ'zent]　　vt. vi. 表现；描绘；代表；回忆；再赠送
inversion　　[ɪn'vɜːʃ(ə)n]　　n. 倒置；倒转；反向反转
scale　　[skeɪl]　　n. 刻度，衡量，数值范围 v. 依比例决定；攀登
flip-flop　　n. 触发器；啪嗒啪嗒的响声　vt. 使翻转；使突然转变
the closed-switch state　开关闭合状态
the open-switch state　开关断开状态
logic gates　[计算机]逻辑闸；逻辑门电路库；逻辑门
AND gate　和门；与门；与电路
binary arithmetic　二进制算术

Notes

[1] The phrase "digital electronics" is used to describe those circuit systems which primarily operate with the use of only two different voltage levels or two other binary states.

译文："数字电子技术"这一术语是用来描述只用两种不同电压电平或两种不同的二进制状态进行工作的电路系统。

说明：句中 to describe... or two other binary states 是不定式短语，充当目的状语。which 引导的定语从句修饰 systems。

[2] In this case, the closed-switch state can be represented by 1 and the open-switch state by 0.

译文：在这种情况下，开关闭合状态用1表示，开关断开状态用0表示。

说明：句中 closed-switch 和 open-switch 具有形容词性，充当定语，均修饰 state。and the open-switch state by 0 已省略了谓语动词部分 can be represented。

[3] An important function of NOT gate is to produce signal inversion or an output signal that is opposite in nature to the input signal.

译文：非门的重要功能是产生反向信号，或者产生与输入信号性质相反的输出信号。

说明：句中不定式 to produce signal inversion or an output signal 充当表语，其后 that 引导的定语从句修饰 an output signal。

[4] ... such as a computer, control or digital-communication system, there are only a few basic operations, which must be performed.

译文：……例如，计算机系统、控制系统或数字通信系统中，需要进行的基本运算也只有几种。

说明：such as 引出 a large scale digital system 的同位语。which must be performed 是非限制性定语从句，对 a few basic operations 起补充说明作用。

Exercises

I. Decide whether the following statements are True (T) or False (F) according to the text.

1. The closed-switch state can be represented by 1 and the open-switch state can be represented by 0.

2. A very common method of digital operation is achieved by using current pulses.

3. Integrated circuits containing many transistors are most commonly used as switching devices in digital electronics logic gates.

4. The three basic types of digital logic gates are the AND gate, the OR gate and the NOT gate.

5. An important function of NOT gate is to produce signal inversion or an input signal that is opposite in nature to the output signal.

6. The three basic types of digital logic gates and the flip-flop are the four circuits most commonly employed in such systems.

II. Translate the following paragraphs into Chinese.

Digital signals and circuits are the vast and important subject. Digital signals are binary in nature taking on values in one of two well-defined ranges. The set of basic operations that be performed on digital signals is quite small. The behavior of any digital system (up to and including the most sophisticated digital computer) can be represented by appropriate combinations of digital variables and the digital operations from this small set.

We concerned with digital system variables that take on only two values (binary variables). We conventionally denote these values as "0" or "1", and then use a special set of rules called Boolean algebra to summarize the various ways in which digital variables can be combined. This algebra and much of the notation are adopted directly from mathematical logic. Thus, "logic variable" or "logic operation" are commonly used in place of digital variable, or digital operation.

Unit 5 Operational Amplifier

Text

The operational amplifier is the most important basic building block of all linear circuits. It has a wide range of applications in such fields as analogue signal operations, amplification, filtering waveforms, producing linear and non-linear signals handling, etc.

The term "operational amplifier" was originally applied to high gain amplifiers operating down to zero frequency which were used in analogue computers to perform certain mathematical operations[1]. These high gain amplifiers are now used for a wide variety of applications, but the name "operational amplifiers" or "op-amp" is normally used even though no mathematical operations are involved. The early operational amplifiers employed discrete components, but it is now much more convenient to employ an integrated circuit. The circuit designers are not generally interested in the internal components of an integrated circuit, but only in the performance of the unit as a whole. Therefore, the symbol in Fig. 5-1 is used to denote the amplifier. It can be seen that there are two inputs, one output and connections to the positive and negative supply lines.

If the inverting input is made slightly more positive, the output will become more negative, this is why the name "inverting" is given to this input. If, however, the non-inverting input is made more positive, the output will also become more positive[2].

Type 741 device is one of the best known general purpose operational amplifiers and is also one of the cheapest of all linear integrated circuits. The device is actually available in a number of different packages. Readers will usually find the type of 741 shown in Fig. 5-2. This type of package is known as the 8 pins dual-in-line.

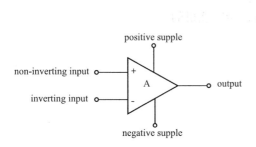

Fig. 5-1 The amplifier

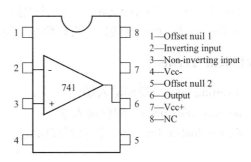

Fig. 5-2 The 8 pins dual-in-line Package of 741

A basic circuit of a 741 inverting amplifier is shown in Fig. 5-3. The input signal is fed to the inverting input of the 741 via R1 and therefore the output is inverted in phase with respect to[3] the input. Operational amplifiers are designed to have a very high input impedance. The input currents of the 741 circuit are therefore very small. As the inverting input is virtually at ground potential, the current which flows through R1 is equal to the input voltage U_i divided by R_3. Similarly, the current

29

flowing through R_3 is $-U_o/R_3$. where U_o is the output voltage. These two currents are almost equal, thus the voltage gain of circuit is equal to the resistor values $G = -R_3/R_1$.

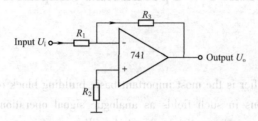

Fig. 5-3 A basic circuit of a 741 inverting amplifier

New Words and Phrases

operational	[ɒpəˈreɪʃ(ə)n(ə)l]	adj.	操作的；运作的，经营的
variety	[vəˈraɪəti]	n.	种类，多样，杂耍变化，多种
discrete	[disˈkriːt]	adj.	离散的，不连续的，n. 分立元件，独立部件
filter	[filtə]	vi.	慢慢传开，滤过，n. 滤波器，筛选，vt. 过滤，渗透
filtering	[ˈfiltəriŋ]	v.	过滤，滤除（filter 的 ing 形式）
integrate	[ˈintigreit]	vt.	使……成整体，vi. 求积分，成为一体，adj. 整合的
denote	[diˈnəut]	vt.	表示，指示，意指
gain	[gein]	n.	收获，增益，利润，vt. 获得，赚到，vi. 获利，增加
respect	[riˈspekt]	n.	尊敬，方面，vt. 尊敬，尊重，遵守
invert	[inˈvəːt]	adj.	转化的，v. 反转，颠倒，反置
package	[ˈpækidʒ]	n.	包，包裹[计]程序包 adj. 一揽子的 vt. 打包；将……包装
pin	[pin]	n.	大头针，钉，引脚，琐碎物，vt. 钉住，将……用针别住
impedance	[imˈpiːdəns]	n.	全电阻，阻抗
in line	adv.	成一直线地，有秩序地，协调地，adj. 联机的	

Technical Terms

inverting input　　反向输入端
non-inverting input　　同向输入端
8 pins dual-in line　　8 管脚双列直插式

Notes

[1] The term "operational amplifier" was originally applied to high gain amplifiers operating down to zero frequency which were used in analogue computers to perform certain mathematical operations.

译文：运算放大器这一术语最早适用于下限频率可低到零赫兹的高增益放大器，这种放大器可在模拟计算机中执行某些数学运算。

说明：句中 which 引导定语从句，修饰前面的 high gain amplifiers，因定语从句并未直接跟在所修饰的先行词后，语法上叫作分隔定语。而定语从句中的 to perform certain mathematical operations 为动词不定式短语，作目的状语。

perform 表示"执行；完成；演奏"，又如：

A series of microcomputer instruction which perform a specific limited task. 执行某一特定有限任务的一系列计算机指令。

［2］If, however, the non-inverting input is made more positive, the output will also become more positive.

译文：然而，如果同相输入端电压正向增大，输出端正电压也将变得更强。

说明：句中 the non-inverting input 表示"同相输入端"，又如：

Inverting input impedance is the same as for the inverting amplifier and the non-inverting input impedance is the sum of R_3 and R_4. 反相输入端的输入阻抗和反相放大器的计算方法相同，同相输入端的输入阻抗为 R_3、R_4 之和。

［3］with respect to 表示"关于；至于"，例如：

With respect to the simulation of the smoke evaluation during a fire, most fire models give the smoke mass density as a result. 火灾中对于浓烟计算的模拟，大部分的火灾模型是以烟质量密度作为模拟结果。

A term used with respect to graphic character to identify type or style ("bold face", "an OCR face"). 一种关于图形字符的术语，用以确定其类型或字体，如黑体、光学字符体。

Exercises

Ⅰ. **Answer the following questions according to the text.**
　　1. Why is the name "operational amplifier" given?
　　2. Can you draw the circuit symbol of an operational amplifier?
　　3. What are the applications of operational amplifiers?

Ⅱ. **Translate the following phrases and expressions.**
　　1. 滤波　　　　　　　　　　2. 运算放大器
　　3. 输入阻抗　　　　　　　　4. 正、负电源端
　　5. 反相输入端虚拟接地　　　6. 作为一个整体的单元性能
　　7. 8 引脚双列直插式集成块　8. 执行各种线性和非线性处理

Ⅲ. **Translate the following sentences, paying more attention to the uses of the underlined parts.**
　　1. The role of a program is to deliver user's intention to a computer.
　　2. The best way to access WWW is with a very fast Web browser.
　　3. To turn on an NPN bipolar transistor, the base must be more positive than the emitter.
　　4. It is necessary for a multimedia system to support a variety of media types.

Ⅳ. **Translate the following passage into Chinese.**
　　Amplifiers are necessary in many types of electronic equipment such as radios, oscilloscopes and record players. Often it is a small alternating voltage that has to be amplified. A junction transistor in the common-emitter mode can act as a voltage amplifier if a suitable resistor, called the

load, is connected in the collector circuit.

The small alternating voltage, the input u_i is applied to the base-emitter circuit and causes small changes of base current which produces large changes in the collector current flowing through the load. The load converts these current changes into voltage changes which form the alternating output voltage u_o.

Translating Skills

增 词 译 法

词的增译就是在译文中增加一些原文中无其形而有其义的词。英语中有时为了避免重复，常省略一些词而不影响全句意义的完整表达，但在汉语译文中如果省略了这些词，就会使译文意义不明确、不通顺。所以，在某些场合的翻译中增词法是非常必要的。

1. 用汉语动词补充英语名词、动名词或介词的意义，使译文通顺

The world needn't be afraid of a possible shortage of coal, oil, natural gas or other sources of fuel for the future. 世界无须担心将来可能出现煤、石油、天然气，或其他燃料来源短缺的问题。

The molecules get closer and closer with the pressure. 随着压力增加，分子越来越接近。

2. 在表达动作意义的英语名词后增添汉语名词

This lack of resistance in very cold metals may become useful in electronic computer. 这种在甚低温中金属没有电阻的现象，可能对电子计算机很有作用。

The lower the frequency is, the greater the refraction of a wave will be. 频率越低，波的折射作用就越强。

3. 增加表示名词复数的词

The moving parts of a machine are often oiled that friction may be greatly reduced. 机器的各个可动部件被润滑油润滑，以便大大减少摩擦。

The first electronic computers used vacuum tubes and others components and this made equipment very large and bulky. 第一代电子计算机使用电子管和其他元器件，这使设备又大又笨。

4. 增加某些被动语态或动名词中没有具体指出的动作执行者或暗含的逻辑主语

The material is said to behave elastically. 人们常说，这种材料具有弹性。

To explore the moon's surface, rockets were launched again and again. 为了勘探月球的表面，人们一次又一次的发射火箭。

5. 在形容词前加名词

According to Newton's Third Law of Motion, action and reaction are equal and opposite. 根据牛顿第三运动定律，作用力和反作用力是大小相等方向相反的。

The washing machine of this type is indeed cheap and fine. 这种类型的洗衣机真是物美价廉。

6. 增加表示数量意义的概括性的词，起修饰润色作用

The frequency, wavelength and speed of sound are closely related. 声音的频率、波长与速度三者密切相关。

A designer must have a good foundation in statics, kinematics dynamics and strength of materials. 一个设计人员必须在静力学、运动学、动力学和材料力学这四个方面有很好的基础。

7. 增加使译文语气连贯的词

In general, all the metals are good conductors, with silver the best and copper the second. 一般来说，金属都是良导体，其中银最佳，铜次之。

Reading

Oscillator

A unit that generates a signal is called an oscillator. Electrical oscillators are widely used in radio and television transmitters and receivers, in signal generators, oscilloscopes and computers, to produce AC with waveforms which may be sinusoidal, square, sawtooth etc. and with frequency from a few hertz up to millions of hertz[1].

Therefore, oscillators may be described in one of two ways in terms of the type of signal waveform generated: sinusoidal oscillators and relaxation oscillators. A sinusoidal oscillator generates a signal having a sine waveform; a relaxation oscillator generates a signal that is usually of square waveform. Sinusoidal oscillators are widely used in radio communication for carrier generation and in many of the test instruments used with such systems. Relaxation oscillators are used as pulse generators in television and radar systems, in digital systems and in test instruments.

Oscillators generally consist of all amplifier and some type of feedback: the output signal is fed back to the input of the amplifier. The frequency-determining elements may be a tuned inductance-capacitance circuit or a vibrating crystal. Crystal controlled oscillators offer the highest precision and stability. Oscillators are used to produce audio and radio signals for a wide variety of purposes. For example, simple audio-frequency oscillators are used in modern push-button telephones to transmit data to the central telephone station for dialing. Audio tones generated by oscillators are also found in alarm clocks, radios, electronic organs, computers, and warning systems. High-frequency oscillators are used in communications equipment to provide tuning and signal-detection[2] functions. Radio and television stations use precise high-frequency oscillators to produce transmitting frequencies.

New Words and Phrases

oscillator ['ɒsɪleɪtə(r)] n. [电子] 振荡器；摆动物；动摇的人
transmitter [trænz'mɪtə] n. 发射机，发报机，传达人
generator ['dʒenəreɪtə] n. 发电机；发生器；生产者
sinusoid ['saɪnəsɔɪd] n. 【数学】正弦曲线 n. 【解剖学】窦状腺，血窦
crystal ['krɪstəl] n. 水晶，晶体，水晶饰品，adj. 水晶的，透明的
precision [prɪ'sɪʒ(ə)n] n. 精度，[数] 精密度；精确 adj. 精密的，精确的
audio ['ɔːdɪəʊ] adj. 声音的；[声] 音频的，声频的
tone [təʊn] n. 音调，语气，色调，vt. 用某种调子说，vi. 颜色调和
tuning ['tjuːnɪŋ] n. 调谐，调音，起弦，协调一致，起音，定音
organ ['ɔːg(ə)n] n. [生物] 器官；机构；风琴；管风琴 n. (Organ) 人名；

33

relaxation oscillators　张弛振荡器
carrier generation　载波

Notes

[1] Electrical oscillators are widely used in radio and television transmitters and receivers, in signal generators, oscilloscopes and computers, to produce AC with waveforms which may be sinusoidal, square, sawtooth etc. and with frequency from a few hertz up to millions of hertz.

译文：电子振荡器广泛应用于无线电和电视信号的发送器和接收器、信号发生器、示波器及计算机中，用以产生正弦波、方波、锯齿波等交流波形，其频率从几赫兹至数百万赫兹。

说明：句中 to produce AC ... 作目的状语，with 短语作后置定语，还有 which 引导的定语从句，均修饰 waveforms。

[2] signal-detection 表示"信号检测"，又如：

The multiple sensor is combined by temperature sensor and photoelectric type smoke sensor, it can realize fire-signal's detection. 复合式探测器由温度探测器、光电感烟探测器构成，以实现火警信号的检测。

Exercises

Ⅰ. **Answer the following questions according to the reading.**
1. What is called an oscillator? Where can oscillators be widely used?
2. How many types of oscillators can be classified? What are they?
3. What are the advantages of crystal controlled oscillators?

Ⅱ. **Translate the following sentences into English.**
1. 正弦波振荡器产生正弦波形的信号。
2. 张弛振荡器产生的通常是方波信号。
3. 高频振荡器用于在通信设备中实现调谐和检波功能。
4. 选频元件可以是电感-电容调谐电路或晶体振荡器。

Unit 6　Heavy and Light Current Engineering

Text

Heavy engineering system needs, in order to function, a source of power. There also needs to be a power distribution network to carry the energy to the various parts of the system. But the purpose for which the power flows in the system is twofold.

1) To power the system so as to enable it to perform the specific tasks.

2) To carry the information around the system, here, energy is the vehicle for conveying information.

Many years ago in electrical engineering we used to differentiate between power engineering (heavy current engineering) and light current engineering.[1] The former was concerned primarily with generation and transportation as well as utilization of electrical power,[2] while the latter covered such aspects as electronics, telephone, radio and the like. The reason for this subdivision was, in part, due to the fact that light current engineering was concerned principally with aspects of communication and control engineering, where only small amounts of power were involved. However, times have changed, and we now realize that we need energy for communication, and the larger the range and the greater the quantity of information to be conveyed, the greater the demands on power[3].

To give a specific illustration, if a teacher speaks to a class of some 10 or 20 students, it's then quite sufficient for him to speak at a normal volume, but if the class is increased to something like 50 or 100 students, he needs to raise his voice and perhaps, at the end of the lecture he might feel tired. When it comes to an even larger audience, perhaps many hundreds or thousands of listeners, then the speaker, in order to make himself heard and to convey the information intended, needs a powerful amplifier to strengthen his voice. Thus we see that the larger the distance and the greater the quantity of information to be conveyed, the greater the demands on power.

Nowadays, a typical radio transmitter has a power of 100 kilowatts so that it can broadcast information over a large area of influence. A similar observation would apply to television and other mass media communications.

We have progressed a long way from the early days of electrical engineering. Nowadays, we have communication systems where the transmitters have powers of thousands of kilowatts in order to achieve the objectives assigned to such apparatus.

Radar is another example. In examining the radar problem it needs to be appreciated that a radar system will usually cover a very large volume of space and will be required to supply an enormous amount of accurate information in a relatively short time. It is because of the multitude and the exactitude of such tasks that the power requirements of a radar system may be many megawatts.[4]

In this way the distinction between heavy current electrical engineering and light current electrical engineering can be said to have disappeared, but we still have the conceptual difference in that in power engineering the primary concern is to transport energy between distant points in space;

while with communication systems the primary objective is to convey, extract and process information, in which process considerable amounts of power may be consumed.

New Words and Phrases

distribution　[ˌdɪstrɪˈbjuːʃn]　n. 分布；分发；分配；散布；销售量
convey　[kənˈveɪ]　vt. 表达；传达；运输；转移
sufficient　[səˈfɪʃnt]　adj. 足够的；充分的
differentiate　[ˌdɪfəˈrenʃieɪt]　vi. 区分，区别　vt. 区分，区别
normal　[ˈnɔːml]　adj. 正常的；正规的；精神健全的　n. 常态；标准
twofold　[ˈtuːfəʊld]　adj. 双重的；两倍的　adv. 双重地；两倍地
assign　[əˈsaɪn]　vt. 分配；指派；[计] [数] 赋值　vi. 将财产过户（尤指过户给债权人）
enormous　[ɪˈnɔːməs]　adj. 巨大的；庞大的
accurate　[ˈækjərət]　adj. 准确的；精确的
illustration　[ˌɪləˈstreɪʃ(ə)n]　n. 说明；插图；例证；图解
appreciate　[əˈpriːʃieɪt]　vt. 欣赏；感激；领会；鉴别　vi. 增值；涨价
multitude　[ˈmʌltɪtjuːd]　n. 群众；多数；大量；众多
observation　[ˌɒbzəˈveɪʃ(ə)n]　n. 观察；监视；观察报告
objective　[əbˈdʒektɪv]　adj. 客观的；目标的；宾格的　n. 目的；目标
megawatt　[ˈmeɡəwɒt]　n. 兆瓦特，百万瓦特
disappear　[ˌdɪsəˈpɪə(r)]　vi. 消失；不见；失踪
consume　[kənˈsjuːm]　v. 消耗；吃喝；毁灭
subdivision　[ˌsʌbdɪˈvɪʒ(ə)n]　n. 细分；分部；供出卖而分成的小块土地
heavy engineering　重型工程，重型机器制造业
power flow　功率通量，能流
electrical engineering　电气工程
heavy current engineering　强电工程
apply to　适用于；运用于
radio transmitter　无线电广播发射机

Notes

[1] Many years ago in electrical engineering we used to differentiate between power engineering (heavy current engineering) and light current engineering.

译文：许多年前，我们在电气工程中常把电力工程（强电工程）和弱电工程区别开来。

说明：句中 used to 是情态动词，没有人称数的变化，这里作为谓语部分，意思是"过去经常"，其中的 to 是不定式符号，所以其后接动词原形（不接动名词）。

[2] The former was concerned primarily with generation and transportation as well as utilization of electrical power.

译文：前者主要同电力的生产、输送和利用有关。

说明：as well as 的含义是"还有""不但……而且……"，可译为"也，又"。值得注意的是，在 A as well as B 的结构里，语意的重点在 A，不在 B，因此，He can speak Spanish as well as English. 的译文应该是："他不但会说英语，而且会讲西班牙语"，绝不能译作"他不但会说西班牙语，而且会讲英语"。如果这样翻译，就是本末倒置了。as well as 和 not only... but also... 同义，但前者的语意重点和后者的语意重点恰好颠倒。

[3] The larger the range and the greater the quantity of information to be conveyed, the greater the demands on power.

译文：距离越远、需要传送的信息量越大，则对电力的需求也随之增加。

说明：句中定冠词 the + 形容词比较级的结构用于表示随着前者的变化，后者呈相应的变化，译为"越……，越……"。

[4] It is because of the multitude and the exactitude of such tasks that the power requirements of a radar system may be many megawatts.

译文：正是由于这些任务的浩繁且要求严密，雷达系统所需要的功率可高达很多兆瓦。

说明：该句为陈述语句型的强调句，其中 because of 可省略。在英语中，陈述句的强调句型为 It is/ was + 被强调部分（通常是主语、宾语或状语）+ that/ who（当强调主语且主语指人）+ 其他部分。译为：正因为……。

Exercises

I. Decide whether the following statements are True (T) or False (F) according to the text.

1. Communication system is used to convey, extract and process information; during this process no power will be consumed.

2. The light current engineering is concerned primarily with generation and transportation as well as utilization of electrical power.

3. The larger the range and the greater the quantity of information to be conveyed, the greater the demands on power.

4. The radar system requires a large quantity of megawatts owing to that the system usually covers a very large volume of space and will be required to supply an enormous amount of accurate information in a relatively short time.

5. The heavy current electrical engineering and light current electrical engineering have no difference in concepts.

II. Match the following phrases in column A with column B.

Column A	Column B
1. heavy current engineering	a. 归因于、归功于
2. a normal volume	b. 无线电广播发射机
3. radio transmitter	c. 正常的体积/大小
4. communication systems	d. 任务的浩繁性严密性
5. multitude and the exactitude of such tasks	e. 大众传媒
6. mass media communications	f. 准确的数据/信息
7. accurate information	g. 通信系统
8. even larger audience	h. 更多观众

9. enormous amount of　　　　　　　i. 大量的
10. due to　　　　　　　　　　　　j. 强电工程

III. Translate the following sentences into English
1. 轮胎里空气越多，里面的压力就越大。
2. 如今一台常见的无线电发射机的功率已达100kW，因此它的播送和影响范围很大。
3. 正是由于这个项目的浩繁且要求严密，两国之间的合作必须要加强。
4. 电气工程自出现以来，已经有了很大的发展。

IV. Translate the following paragraph into Chinese.

Nowadays, the electrical engineering has been widely applied in our daily life since it first started. For example, a typical radio transmitter can broadcast information over a large area of influence. Another common application is radar system and similar observation would come to television and other mass media communications.

Translating Skills

减 词 译 法

减词翻译法又称为省略法或省译法，它是指在译成汉语时，把原文中的某个（些）词不译出来。减词翻译法是科技英语翻译常用的重要技巧。

1. 代词的省译
代词的省译主要表现在人称代词、反身代词和物主代词等几方面。
We should concern ourselves here only with the structure of atoms. 这里我们只论述原子的结构。（省译反身代词）
It is now clear why the laser usually oscillates on the R_1 line. 现在清楚了，为什么激光总是在 R_1 线上振荡。（省译形式主语 it）

2. 冠词的省译
冠词的省译主要有定冠词、不定冠词和固定词组中的冠词的省译。
The thinner the air, the less support it gives to the plane. 空气越稀薄，给飞机的支撑力就越小。（省译定冠词）
Satellites can be sent into space with the help of rockets. 借助火箭可以把卫星送入太空。（省译定冠词）

注意：在常用的英语词组中，往往因冠词使词组的意义发生了很大改变。所以，在翻译时虽不必译出冠词本身的意义，但要重视冠词在整个词组中所起的改变意思的作用。例如：
In the past, to fly to the moon was out of the question. 过去，飞往月球是绝对做不到的。
In these days, to fly to the moon is out of question. 现在，飞往月球是不成问题的。

3. 动词的省译
英语的动词一般是不能省译的。但某些动词在某种场合又是可以省译的。例如：
This laser beam covers a very narrow range of frequency. 这个激光束的频率范围很窄。（省译动词 covers）
Then came the development of the microcomputer. 后来，微型计算机发展起来了。（省译

动词 came）

4. 介词的省译

介词的省译是因为汉语中的介词用得较少，词与词之间的意思往往通过上下文表示出来。例如：

Electronic computers of various types are produced in our country. 我国生产各种类型的计算机。（省译表示地点的介词 in）

The research <u>for</u> even better magnetic material is a part of the modern frontier of physics. 研制更好的磁性材料是现代物理学前沿之一。（省译做定语的介词 for）

5. 连词的省译

This machine has worked in succession for seven <u>or</u> eight hours. 这台机器已经连续运转了七、八个小时。（省译并列连词 or）

All of us know <u>that</u> the conductivity of semiconductors changes with temperature. 我们都知道，半导体的导电性随温度而变化。（省译引导宾语从句的连词 that）

6. 其他词语的省译

<u>Rubbing, or friction</u>, gives very little electricity. 摩擦只能产生少量静电。（省略相同意义的重复词）

As we know, electrons revolve about the nucleus, or center, of an atom. 正如我们所知，电子围绕着原子核旋转。（省略同位语）

Reading

Superconductivity

In the normal conduction of an electric current by a metal, electrons are pumped through the crystal lattices of the metal under the influence of an electromotive force supplied by a cell, battery or a generator. The motion of the electrons is impeded by collisions with the atoms in the lattice; as we have seen, this impedance is responsible for the conductor's electrical resistance. The resistance increases as the temperature increases, because the greater vibrations of the atoms interfere with the electrons' motion more strongly.

Suppose that if the atoms vibrations are completely stilled by reducing the temperature to absolute zero, the electrical resistance might disappear. To test the idea, Heike Kamerlingh-Onnes passed an electric current through some frozen mercury and discovered that all resistance to the flow of current disappeared at 4.15K. This phenomenon — called superconductivity — occurred at several degrees above absolute zero, when the resistance of mercury disappeared.

Superconductivity is not at all a rare phenomenon. [1] 26 elements and several hundred compounds and alloys are already known to be superconductors. In each case the electrical resistance disappears at a particular temperature, that is, 0.01K for tungsten, 20.7K for an alloy of aluminum, germanium, and niobium. It is not necessary to go to the absolute zero.

In the superconducting state, a stream of electrons can flow without encountering any resistance. Such currents, once started, continue flowing for long periods with little or no energy. In one experiment, an electric current induced in a small ring of metal was still circling a year later, and had not decreased in strength.

Practical applications of superconductivity have been made only recently. In most cases it is necessary to use liquid helium to reach the low temperatures required; it was not until the 1960s that helium became available at a price that made superconducting devices practical.

Superconducting magnets are the largest class of devices in use, and large electric motors using superconductors have been built. They have been proposed for trains, which could travel 300 miles per hour by eliminating the problem of friction; a train equipped with such a motor would, in effect, be lifted off the tracks by using the force between current elements.

Superconductors have also been proposed for amplifiers, particle accelerators and computers.

New Words and Phrases

  generator   ['dʒenəreitə(r)]   n. 发电机；发生器
  impede   [im'pi:d]   vt. 妨碍；阻止
  mercury   ['məkjəri]   n. [化] 汞，水银（M~）水星；精神，元气
  induce   [in'dju:s]   vt. 引起；促使；劝服；引诱
  equip   [i'kwip]   vt. 装备；具备；准备
  helium   ['hi:liəm]   n. [化学] 氦（符号为 He，2号元素）
  lattice   ['lætɪs]   n. [晶体] 晶格；格子；格架 vt. 使成格子状
  a stream of   一股
  liquid helium   液态氦
  crystal lattice   晶体点阵，一种由晶体的原子、分子或离子在空间各点所作的几何；排列也作 "space lattice"。
  absolute zero   绝对零度，在此温度下物质没有热能，相当于摄氏 -273.15 度或华氏 -459.67度。

Notes

[1] Superconductivity is not at all a rare phenomenon.
译文：超导性是一种很普遍的现象。
说明：not at all 用来表示否定（是 no 的加强说法），意为"一点也不；完全不"。rare 作为形容词是"稀有的"，这里是用形容词来表达否定，同样的形容词还有 few, little, scare 等。前后两个否定构成双重否定，加强了句子肯定的语气。

Exercises

I. Answer the following questions according to the passage.
 1. Why does the resistance increases as the temperature increases?
 2. What phenomenon is called superconductivity?
 3. What are the practical applications of superconductivity?

Ⅱ. **Translate the follouing sentences into English.**
1. 超导性并不是罕见的现象。
2. 假定把温度降低到绝对零度,使原子振动完全停止,电阻就会消失。
3. 在超导状态下,电流的流动不会遇到任何阻力。
4. 众所周知,这种阻碍是导体产生电阻的原因。

Unit 7 How Power Grids Work

Text

Electrical power is a little bit like the air you breathe: you don't really think about it until it is missing. [1] Power is just "there", meeting your every need, constantly. It is only during a power failure, when you walk into a dark room and instinctively hit the useless light switch that you realize how important power is in your daily life. [2] You use it for heating, cooling, cooking, refrigeration, light, computation, entertainment. Without it, life can get somewhat cumbersome.

Power travels from the power plant to your house (Seen in Fig. 7-1) through an amazing system called the power distribution grid. The grid is quite public and now we will look at all of the equipment that brings electrical power to your home.

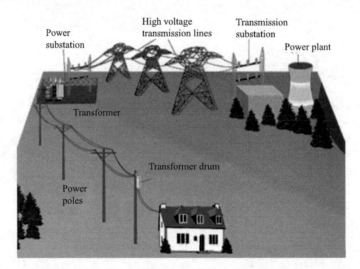

Fig. 7-1 Power travels from the power plant to your house

1. The Power Plant

Electrical power starts at the power plant. In almost all cases, the power plant consists of a spinning electrical generator. Something has to spin that generator—it might be a water wheel in a hydroelectric dam, a large diesel engine or a gas turbine. But in most cases, the thing spinning the generator is a steam turbine. The steam might be created by burning coal, oil or natural gas, or the steam may come from a nuclear reactor. No matter what it is that spins the generator, commercial electrical generators of any size generate what is called three-phase AC power. To understand three-phase AC power, it is helpful to understand single-phase power first.

2. Alternating Current

Single-phase power is what you have in your house. You generally talk about household electrical service as single-phase, 220-Volt AC service. If you use an oscilloscope and look at the power found at a normal wallplate outlet in your house, what you will find is that the power at the wall plate looks like a sine wave, and that wave oscillates between −311 Volts and 311 Volts (the peaks are indeed at 311 Volts; it is the effective voltage that is 220 Volts). The rate of oscillation for the sine wave is 50 cycles per second. Oscillating power like this is generally referred to as AC, or alternating current.

3. The Power Plant Produces AC

The power plant produces three different phases of AC power simultaneously, and the three phases are offset 120° from each other. There are four wires coming out of every power plant: the three phases plus a neutral or ground common to all three. If you were to look at the three phases on a graph, they would look like this (Seen in Fig. 7-2) relative to ground.

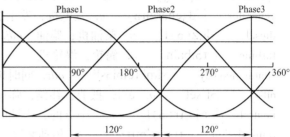

Fig. 7-2 The waveform of 3-phase AC

4. Transmission Substation

The three-phase power leaves The generator and enters a transmission substation at the power plant. This substation uses large transformers to convert the generator's voltage (which is at the thousands of volts level) up to extremely high voltages for long-distance transmission on the transmission grid.

5. The Distribution Grid

For power to be useful in a home or business, it comes off the transmission grid and is stepped-down to the distribution grid. This may happen in several phases. The place where the conversion from "transmission" to "distribution" occurs is in a power substation. [3]

6. Distribution Bus

The power goes from the transformer to the distribution bus. In this case, the bus distributes power to two separate sets of distribution lines at two different voltages. The smaller transformers attached to the bus are stepping the power down to standard line voltage (7200V) for one set of lines, while power leaves in the other direction at the higher voltage of the main transformer. The power leaves this substation in two sets of three wires, each headed down the road in a different direction.

The three wires at the top of the poles are the three wires for the three-phase power. The fourth wire lower on the poles is the ground wire. In some cases there will be additional wires, typically phone or cable TV lines riding on the same poles.

7. Taps and at the House

A house needs only one of the three phases, so typically you will see three wires running down a main road, and taps for one or two of the phases running off on side streets. The transformer's job is to reduce the 7200 Volts down to the 220 Volts that makes up normal household electrical service. The 220 Volts enters your house through a typical watt-hour meter.

New Words and Phrases

instinctively	[in'stiŋktivli]	adv. 本能地
refrigeration	[rifri:dʒ'reiʃn]	n. 制冷；冷藏；冷却
computation	[kɔmpju'teiʃən]	n. 估计，计算
cumbersome	['kʌmbəs(ə)m]	adj. 笨重的；累赘的；难处理的
spinning	['spiniŋ]	n. 纺纱 adj. 纺织的 v. 旋转，纺织；纺纱；晕眩；结网
generator	['dʒenəreitə]	n. 发电机；发生器；生产者
diesel	['di:z(ə)l]	n. 柴油机；柴油 adj. 内燃机传动的；供内燃机用的
turbine	['tə:bain]	n. 涡轮；涡轮机
simultaneously	[ˌsiml'teiniəsli]	adv. 同时地
offset	[ɔf'set]	n. & v. 抵消，补偿，弥补
neutral	['nju:tr(ə)l]	adj. 中立的，中性的
transmission	[træns'miʃən]	n. 传动装置，变压器；传递；传送
substation	['sʌbsteiʃ(ə)n]	n. 分局；变电所
phase	[feiz]	n. 相；阶段

hydroelectric dam　水电坝
gas turbine　燃气轮机；燃气涡轮
nuclear reactor　核反应堆；原子炉
transmission grid　输电网
watt-hour meter　电表
power grid　电力网
power plant　发电厂

Notes

［1］Electrical power is a little bit like the air you breathe: you don't really think about it until it is missing.

译文：电力网有点像你呼吸的空气，除非没有了否则你不会想到它。

说明：句中"not. . . until"意思是"直到……才……"，例如：

The old man did not get off the bus until it stopped.

［2］It is only during a power failure, when you walk into a dark room and instinctively hit the useless light switch that you realize how important power is in your daily life.

译文：只有当停电时，当你在黑暗的房间里走动本能地想去打开无用的开关时，你才会发现电在你的日常生活中是多么重要。

说明：本句中"It is. . . that. . ."为强调句型。该结构不能用来强调谓语动词、定语和表语，可强调主语、宾语和状语，其句型为 It is/was + 强调成分 + that + 其他成分，若强调成分是人，也可用 who 代替 that。例如：

It was in the library that he read the book English for Automation yesterday afternoon.

［3］The place where the conversion from "transmission" to "distribution" occurs is in a power substation.

译文：把电从输电网转换到配电网的地方是变配电站。

说明：句中 where 引导定语从句，修饰先行词 place。例如：

Beijing is the place where I was born.

Exercises

I. Answer the following questions according to the passage.

1. How does power travel from the power plant to our house?
2. What does the power plant produce?
3. Where does the conversion from "transmission" to "distribution" occur?

II. Fill in the blanks.

1. You use power for heating, _____, cooking, refrigeration, light, _____, entertainment.
2. The steam might be created by burning coal, oil or _____, or the steam may come from a _____ reactor.
3. The three-phase power leaves the _____ and enters a _____ at the power plant.
4. At each house, there is a _____ attached to the pole.

III. Fill in the blanks with the proper word or phrase. Change the form if necessary.

damage	security	limit	minimize
make up	maintain	transmission	operate

Up until now we have been mainly concerned with _____ the cost of operating a power system. An overriding factor in the _____ of a power system is the desire to maintain system _____. For example, a generating unit may have to be taken off-line because of auxiliary equipment failure. By _____ proper amounts of spinning reserve, the remaining units on the system can _____ the deficit without too low a frequency drop or need to shed any load. Similarly, a transmission line may be _____ by a storm and taken out by automatic relaying. If in committing and dispatching generation proper regard for transmission flows is maintained, the remaining _____ lines can take the increased loading and still remain within _____.

IV. Translate the following paragraph into Chinese.

Because electricity can not be massively stored under a simple and economic way, the production and consumption of electricity must be done simultaneously. A fault or misoperation in any stages of a power system may possibly result in interruption of electricity supply to the customers. Therefore, a normal continuous operation of the power system to provide a reliable power supply to the customers is of paramount importance.

Translating Skills

科技英语词汇的结构特征（Ⅰ）

科技英语词汇来源如下：

1）来自英语中的普通词汇，但在科技英语中被赋予了新的词义。例如：Work is the transfer of energy expressed as the product of a force and the distance through which its point of application moves in the direction of the force. 在这句话中，"work" "energy" "product" "force" 都是从普通词汇中借来的物理学术语。"work" 的意思不是 "工作"，而是 "功"；"energy" 的意思不是 "活力"，而是 "能"；"product" 的意思不是 "产品"，而是 "乘积"；"force" 的意思不是 "力量"，而是 "力"。

2）来自希腊或拉丁语等语言。

例如：therm—热（希腊语），thesis—论文（希腊语），parameter—参数（拉丁语），radius—半径（拉丁语）。

这些来源于希腊语和拉丁语的词的复数形式有些仍按原来的形式，如 thesis 的复数形式是 theses，stratus 的复数形式是 strati；但由于在英语里使用时间较长，不少词除保留了原来的复数形式外，又采用了英语的复数形式，例如：formula（公式，拉丁语）的复数形式可以是 formulae，也可以是 formulas，stratum 的复数形式（层，拉丁语）可以是 strata，也可以是 stratums。

3）新出现的词汇。每当出现新的科学技术现象时，人们都要通过词汇把它表示出来，这就需要构造新的词汇。构造新词主要有以下几种方法。

① 转化：通过词类转化构成新词。英语中名词、形容词、副词、介词可以转化成动词，形容词、副词、介词可以转化成名词；但最活跃的是名词转化成动词和动词转化成名词。例如，名词 island（小岛）转化成动词 island（隔离），动词 coordinate（协调）转化成名词 coordinate（坐标）。

② 合成：由两个独立的词合成为一个词。例如：air + craft—aircraft（飞机），air + port—airport（机场），metal + work—metalwork（金属制品），power + plant—powerplant（发电站）。

有的合成词的两个成分之间要有连字符，例如，cast-iron（铸铁），conveyer-belt（传送带），machine-made（机制的）。英语中有很多专业术语由两个或更多的词组成，叫作复合术语。它们的构成成分虽然看起来是独立的，但实际上合起来构成一个完整的概念；因此应该把它们看成是一个术语，如：liquid crystal—液晶，water jacket—水套，computer language—计算机语言。

③ 派生：这种方法也叫缀合。派生词是由词根加上前缀或后缀构成的。构成新词，只改变词义，不改变词性。例如：

　　decontrol（取消控制）　　　　　v. de + control（control 是动词）；
　　ultrasonic（超声的）　　　　　　adj. ultra + sonic（sonic 是形容词）；
　　subsystem（子系统）　　　　　　n. sub + system（system 是名词）。

有些加前缀的派生词在前缀和词根之间有连字符，例如，hydro-electric（水电），extra-heavy（超重）。英语的前缀是有固定意义的，记住其中的一些常用前缀对于记忆生词和猜测词义有帮助。这里只举一些表示否定的前缀来说明问题，例如：anti-表示"反对"，antibody（抗体）；counter-表示"反对，相反"，counterbalance（反平衡）；contra-表示"反对，相反"，contradiction（矛盾）；de-表示"减少，降低，否定，去除"，decrease（减少），devalue（贬值），decode（解码）；dis-表示"否定，除去"，discharge（放电），disassemble（拆卸）；in，il（单词首字母为l）表示"不"，inaccurate（不准确的）；im-（在字母m、b、p前）表示"不"，imbalance（不平衡的），impure（不纯的）；mis-表示"错误"，mislead（误导）；non-表示"不，非"，non-ferrous（有色金属的）；un-表示"不，未，丧失"，unaccountable（说明不了的），unknown（未知的），unbar（清除障碍）。

　　加后缀构成新词可能改变也可能不改变词义，但一般改变词性。例如：
　　electricity n. electric + ity（electric 是形容词）；
　　liquidize（液化）v. liquid + ize（liquid 是名词）；
　　conductor n. conduct + or（conduct 是动词）；
　　invention n. invent + ion（invent 是动词）。
　　有的派生词加后缀的时候，语音或拼写可能发生变化。例如：
　　simplicity（单纯）n. simple + icity；
　　maintenance（维修）n. maintain + ance；
　　propeller（推进器）n. propel + l + er。
　　英语后缀的作用和前缀有所不同，它们主要用来改变词性。从一个词的后缀可识别它的词性，这是它的语法意义。它们的词汇意义往往并不明显，下面是一些常见的形容词后缀。例如：
　　-able，-ible，-uble 表示"可……的"，avoidable（可以避免的），audible（听得见的），soluble（可溶的）；
　　-al，-ant，ent，表示属性、性质，译为"……的"，fundamental（基本的），abundant（富饶的），apparent（显然的）；
　　-ed，表示"有……的"，cultured（有文化的）；
　　-ful，表示"充满……的，有……倾向的"，useful（有用的）；
　　-ic，ieal，形容词尾，basic（基本的），economical（经济的）；
　　-less，表示"没有"，useless（无用的）；
　　-ous，形容词尾，numerous（众多的）；
　　-let，表示"小的"，booklet（小册子），wavelet（小波）。

Reading

Transformers

　　Transformers come in many sizes. Some power transformers are as big as a house. Electronic transformers, on the other hand, can be as small as a cube of sugar. All transformers have at least one coil; most have two although they may have many more.

　　The usual purpose of transformers is to change the level of voltage. But sometimes they are used to isolate a load from the power source.

1. Types of Transformers

Standard power transformers have two coils. These coils are labeled primary and secondary. The primary coil is the one connected to the source. The secondary is the one connected to the load. There is no electrical connection between the primary and secondary. The secondary gets its voltage by induction.

The only place where you will see a step-up transformer is at the generating station. Typically, electricity is generated at 13,800 Volts. It is stepped up to 345,000 Volts for transmission. The next stop is the substation where it is stepped down to distribution levels, around 15,000 Volts. Large substation transformers have cooling fins to keep them from overheating. Other transformers are located near points where the electric power is used.

2. Transformer Construction

The coils of a transformer are electrically insulated from each other. There is a magnetic link, however. The two coils are wound on the same core. Current in the primary magnetizes the core. This produces a magnetic field in the core. The core field then affects current in both primary and secondary.

There are two main designs for cores.

1) The core type has the core inside the windings.
2) The shell type has the core outside.

Smaller power transformers are usually of the core type. The very large transformers are of the shell type. There is no difference in their operation, however.

Coils are wound with copper wire. The resistance is kept as low as possible to keep losses low.

3. Idealized Transformers

Transformers are very efficient. The losses are often less than 3 percent. This allows us to assume that they are perfect in many computations.

Perfect means that the wire has no resistance. It also means that there are no power losses in the core. Further, we assume that there is no flux leakage. That is, all of the magnetic flux links all of the turns on each coil.

4. Excitation Current

To get an idea of just how small the losses are, we can take a look at the excitation current. Assume that nothing is connected to the secondary. If you apply rated voltage to the primary, a small current flows. Typically, this excitation current is less than 3 percent of rated current.

Excitation current is made up of two parts. One part is in phase with the voltage. This is the current that supplies the power lost in the core. Core losses are due to eddy currents and hysteresis.

Eddy currents circulating in the core result from induction. The core is, after all, a conductor within a changing magnetic field.

Hysteresis loss is caused by the energy used in lining up magnetic domains in the core. The alignment goes on continuously, first in one direction, then in the other.

The other part of the excitation current magnetizes the core. It is this magnetizing current that supplies the "shuttle power". Shuttle power is power stored in the magnetic field and returned to the source twice each cycle. Magnetizing current is quadrature (90° out of phase) with the applied voltage.

New Words and Phrases

transformer　[trænsˈfɔːmə]　*n.* 变压器
substation　[ˈsʌbsteiʃn]　*n.* 变电所
induction　[inˈdʌkʃ(ə)n]　*n.* （电磁）感应
fin　[fin]　*n.* 散热片
magnetize　[ˈmægnitaiz]　*vt.* 使磁化；吸引
flux　[flʌks]　*n.* 流量；变迁；不稳定；流出
leakage　[ˈliːkidʒ]　*n.* 漏；漏损物；泄漏；漏损量
hysteresis　[ˌhistəˈriːsis]　*n.* 滞后作用，磁滞现象
excitation　[eksiˈteiʃn]　*n.* 激发；励磁；刺激
shuttle　[ʃʌtl]　*n.* 梭形；往复
quadrature　[ˈkwɔdrətʃə]　*n.* 求积；矩；上（下）弦
electronic transformer　电力变压器
step-up transformer　升压变压器
excitation current　励磁电流
in phase with　与……相同
eddy current　涡轮电流
shuttle power　往复能量

Exercises

Ⅰ. **Answer the following questions according to the text.**
1. What is the usual purpose of transformers?
2. Why do large substation transformers have cooling fins?
3. What is the shuttle power?

Ⅱ. **Translate the following sentences into English.**
1. 变压器的线圈之间是绝缘的，但存在磁联系。
2. 铁心磁场影响一次和二次绕组的电流。
3. 欲知损耗的大小，可以看一下励磁电流。

49

Unit 8　Basic Relay Types

Text

　　There are several possible ways to classify relays: by function, by construction and by application. Relays are one of two basic types of construction: electromagnetic or solid state. The electromagnetic type relies on the development of electromagnetic forces on movable members, which provide switching action by physically opening or closing sets of contacts. The solid state variety provides switching action with no physical motion by changing the state of serially connected solid state component from nonconducting to conducting (or vice versa). Electromagnetic relays are older and more widely used; solid state relays are more versatile, potentially more reliable, and fast.

　　In structure, relays can be classified as follows:

　　1) Moving-Armature Relay. The moving-armature relay is shown in Fig. 8-1a. A coil C is wound on a core of magnetic material M. A movable armature A carries with it a movable contact B. With no current in the coil, the contact B rests against the contacts 1-1, connecting them together. Gravity or spring holds the armature in the position shown. If current through the coil is slowly increased, a point is reached at which the armature A suddenly rotates to the left, opening contacts 1-1 and closing contacts 2-2.[1] The minimum current at which this motion may be initiated is termed the pickup current. If the current is now slowly reduced, a value will be found at which the armature will drop back to its original position. This is termed the drop-out or release current.

　　2) Solenoid Relay. A solenoid relay is shown in Fig. 8-1b. A fixed coil surrounds a movable iron plunger. The plunger is held away from a symmetrical position in the coil by gravity or by a spring. When current of sufficient value flows in the coil, the iron core is drawn into the coil, thus closing the contacts by a simple position change. Again the pickup current is greater than the release current. The plunger in any selected position experiences a force proportional to the square of the current (or to the square of the voltage across the coil).

　　3) Balance Beam. An iron armature is fastened to a balance beam as shown in Fig. 8-1c. The armature is drawn into a fixed-position coil when current flows in the coil. Various contact and spring arrangements are possible.

　　4) Mutually Interacting Coils Relay. Two coils, one fixed and one movable, are positioned so that their magnetic fields interact (Seen in Fig. 8-1d). If the currents are of constant direction, the force between them will be proportional to the algebraic product of the two currents and may result in drawing the coils together or thrusting them apart. If AC currents flow in the coils, the force action will be proportional to the product of the two current values and the cosine of the angle Δ between them:

$$F = K_M I_1 I_2 \cos\Delta$$

　　5) Induction Disc or Induction Cup Relay. The induction-disc relay operates on the principle of the induction motor. As illustrated with Fig. 8-1e, currents displaced in time flowing through coils displaced in space give rise to a moving magnetic field which passes through the movable disc. This moving magnetic field sets up within the disc currents which result in force action on the disc, tend-

ing to make it rotate.[2] In this type of relay the coils above the disc may carry one current and the coils below the disc carry another current independently. If these currents are displaced from each other by the angle Δ, the torque produced on the disc is

$$T = K_D I_1 I_2 \sin\Delta$$

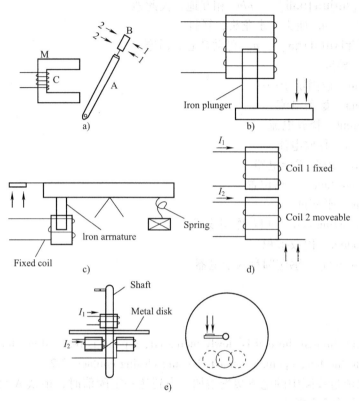

Fig. 8-1 Electromechanical relay types
a) Moving armature b) Solenoid c) Balance beam
d) Mutually interacting coils e) Induction disc, elevation and plan

If the two sets of coils are in series and so carry the same current, no torque results. However, a resistor or a capacitor in parallel with one set of coils causes its current to shift out of phase with the other and so gives rise to torque. Other methods of phase shifting are used. The coils of induction relays are usually provided with taps which enable the selection of pickup current. This permits proper protection coordination even after changes in load layout have taken place during years of service.

New Words and Phrases

serially   [ˈsiəriəli]   adv. 逐次地，连载地，串行地
contact   [ˈkɒntækt]   n. 接触器，电气接头，触头
versatile   [ˈvɜːsətaɪl]   adj. 多用途的，多功能的
spring   [sprɪŋ]   n. 弹簧，发条

armature	[ˈɑːmətʃə]	n. 电枢，转子，电枢线圈，衔铁线圈
rotate	[rə(ʊ)ˈteɪt]	v. 旋转，转动
torque	[tɔːk]	n. 力矩，转矩
layout	[ˈleɪaʊt]	n. 规划，编排，布局
mutually	[ˈmjuːtʃuəli]	adv. 相互地，彼此地
tap	[tæp]	n. 抽头，水龙头，轻打
plunger	[ˈplʌn(d)ʒə]	n. 可动铁心，活塞

give rise to　导致
be fastened to　被固定于……
pickup current　始动电流
drop-out current　释放电流
release current　释放电流
solenoid relay　螺管式继电器
symmetrical position　平衡位置
balance beam　平衡杆
mutually interacting coils　相互作用线圈
algebraic product　代数乘积
induction-disc relay　感应圆盘式继电器

Notes

[1] If current through the coil is slowly increased, a point is reached at which the armature A suddenly rotates to the left, opening contacts 1-1 and closing contacts 2-2.

译文：如果流过线圈中的电流缓慢增加，当到达一定阈值时，衔铁 A 会突然旋向左边，断开 1-1 触点并闭合 2-2 触点。

说明：句中有一个介词＋关系代词构成的定语从句 at which the armature A suddenly rotates to the left，处于句子结构上的需要，该从句与其被修饰词 a point 之间隔着主句谓语 is reached。从句之后的两个并列的现在分词短语 opening contacts 1-1 and closing contacts 2-2 可理解为定语从句的状语，表示结果。

[2] As illustrated with Fig. 8-1e, currents displaced in time flowing through coils displaced in space give rise to a moving magnetic field which passes through the movable disc. This moving magnetic field sets up within the disc currents which result in force action on the disc, tending to make it rotate.

译文：如图 8-1e 所示，时变电流流经空间位置变化的线圈导致移动磁场，该磁场穿过可动圆盘，在圆盘中激发出感生电流，（这个感生电流在磁场中）会使圆盘受到力的作用，带动圆盘旋转。

说明：这段文字由两个逻辑关系十分紧密的句子组成，不可将它们分开理解。

在第一个句子的主语 currents 之后有两个修饰成分，一个是过去分词短语 displaced in time，另一个是现在分词短语 flowing through coils displaced in space。coils 之后的过去分词短语 displaced in space 是 coils 的定语。

第二个句子中的定语从句 which result in force action on the disc, tending to make it rotate

修饰 the disc currents。从句后的分词短语 tending to make it rotate 可看作从句的状语，表示结果。

Exercises

Ⅰ. **Answer the following questions according to the text.**

1. What's the difference between an electromagnetic relay and a solid state relay?
2. How does the moving-armature relay work?
3. How many kinds of relays are there by classifying them in structure? And what are they?
4. What's the relationship between the force that the plunger experiences and the current in a solenoid relay?

Ⅱ. **Decide whether the following statements are true（T）or False（F）according to the text.**

1. The solid state relies on the development of electromagnetic forces on movable members, which provide switching action by physically opening or closing sets of contacts.
2. The plunger of a solenoid relay in any selected position experiences a force proportional to the voltage across the coil.
3. When a moving-armature relay works, the minimum current at which this motion may be initiated is termed the release current. If the current is now slowly reduced, a value will be found at which the armature will drop back to its original position. This is termed the pickup value.
4. When a Mutually Interacting Coil works, if direct current flows in the coils, the force action will be proportional to the algebraic product of the wo currents.
5. Various contact and spring arrangements can be used in the balance beam relays.

Ⅲ. **Translate the following terms into English.**

1. 电磁继电器　　　　　　　2. 固态继电器
3. 可动衔铁式继电器　　　　4. 释放电流
5. 始动电流　　　　　　　　6. 螺线管式继电器
7. 平衡杆式继电器　　　　　8. 相互作用线圈
9. 感应圆盘式继电器　　　　10. 可动圆盘

Ⅳ. **Translate the following sentences into English.**

1. 一切物体都能在某种程度上传热，但许多固体比液体传热性能好，而液体又比气体好。
2. 行星在围绕太阳旋转的同时，也绕着它们的自转轴自转。
3. 电流与电压成正比，与电阻成反比。
4. 这些恶劣的环境已造成了许多犯罪行为。

Translating Skills

科技英语词汇的结构特征（Ⅱ）

缩略：把词省略或简化，然后组合成新词。现在的趋势是缩略词的数目不断增大，使用面不断扩大。例如：

laboratory 截短成 lab（实验室）；

unidentified flying object 缩合成 UFO（不明飞行物）；

radio detection and ranging 缩减成 radar（雷达）；

transistor 和 receiver 各取一部分，组合成 transceiver（收发机）；

motor + hotel→motel（汽车旅店）；

mechanics + electronics→mechatronics（机电学）；

telecommunications satellite 缩略为 Telesat（通信卫星）。

英语科技文体中有很多词汇并不是专业术语，且在日常口语中用得不是很多，它们多见于书面语中，叫作文语词。掌握这类词对阅读科技文献或写科技论文十分重要。这类词用得很广，它们不仅出现在各专业的科技文体中，也出现在政治、经济、法律、语言等社会科学的文体中，例如：

accordance（按照），imply（隐含），acknowledge（承认），inclusion（包括），alter（改变），incur（招致），alternative（交替的），indicate（指示），amend（修正），induce（归纳），application（应用），initial（初始的），appropriate（恰当的），modification（修改），attain（达到），nevertheless（然而），circumstance（情况），nonetheless（然而），compensation（补偿），obtain（获得），confirm（证实），occur（发生），consequence（后果），omission（省略），considering（鉴于），providing（假设），consist（组成），reduce（减少），constitute（构成），replacement（代替），consume（消耗），specifically（具体地说），deduce（推理），distinction（差别），valid（有效的），function（功能），verify（验证），illustrate（说明）。

关于缩略词的大量采用

采用的缩略词有以下几种：

1）缩略词的首部（front clipping），如 telephone = phone，university = varsity，helicopter = copter，etc.

2）缩略词的尾部（back clipping），如 advertisement = ad，debutante = deb，modern = mod，professional = pro，exposition = expo，memorandum = memo，etc.

3）缩略词的首尾部，如 influenza = flu，detective = tec，refrigerator = fridge，etc.

4）首字母缩略词，如 V. O. A. = Voice of America，IOC = International Olympic Committee，WTO = World Trade Organization，CAD = Computer Aided Design，NATO = Noah Atlantic Treaty Organization，CAE = Computer Aided Engineering，CAM = Computer Aided Manufacture，etc.

约定的专业术语：

在科技英语中大量出现专业术语，专业术语语义较固定，翻译也是固定表达方式，如 TCP/IP（Transmission Control Protocol/Internet Protocol，传输控制协议/国际网协议），relay（继电器），interface（计算机通信接口），point-to-point（点对点），topology（拓扑学），HTML（Hypertext Markup Language，超文本标识语言），HTTP（Hypertext Transfer Protocol，超文本传输协议），matrix（矩阵），prototype（原型，样机），state-of-the-art（工艺水平），detect（探测，检测），fault（故障），hydraulic system（液压系统），NC（Numerical Control，数控），milling（铣，磨），feed（进给量），machine tool（机床），feedback（反馈），hardness（硬度），coolant（冷却液），roughness（粗糙度），sensor（传感器），integration（集成，综合），pulse train（脉冲序列），patent（专利），poly-phase（多相的），transformer（变压器），integrate circuit (集成电路)。

Reading

Fault-Clearing Protective Relays

Overcurrent relaying slow-speed relays. The most obvious effect of a fault is to change the current in the faulted conductor from a normal value to an abnormally large one. Therefore it is not surprising that the earliest methods of clearing faults were based on the utilization of that effect (overcurrent). Early methods included fuses, circuit breakers with series trip coils, and slow-speed overcurrent relays.

Slow-speed overcurrent relays are mostly of the induction type. To obtain selectivity without unnecessarily long delay, such relays usually have a delay which varies inversely with the current. Both time and current settings are adjustable. Since the fault current decreases, on account of the increased impedance of the line between the fault and the source, as the fault is moved farther from the source of power, it follows that the relay operating time increases as the distance to the fault increases.

The time-distance curves change with such conditions as connected generating capacity and the connection or disconnection of other transmission lines, and therefore, to ensure selectivity, curves should be checked for several conditions to ascertain that, under the worst condition, an adequate interval exists between the operating times of relays, and similarly, between each pair of relays on adjoining line sections. Coordination may be accomplished by judicious choice of both time settings and current settings.

If the relay current changes but little with fault location, the curve of relay time versus fault position becomes more like curve a than like curve b. Such a condition is likely to exist if the impedance of the protected section is small compared with the impedance between the generators and the protected section, as may well be true if the section is short and is fed solely or principally from one end. Moreover curve a may hold even though the relay current does change with fault location, if, as is usually true, the relays are operating on the minimum-time part of their time-current characteristic. Curve a represents an undesirable condition when several protected line sections are in cascade, because the relay time of the lines near the source of power becomes increasingly long.

If a line section is long or has power sources at each end, the relay current will vary considerably with fault location. But, even if the current varies enough to give a curve like b, the operating time of a relay near the generator is usually somewhat longer than that of a relay farther from the generator, though not so much longer as in curve a.

By the use of graded time settings, overcurrent relays can always be made to work selectively on a radial transmission or distribution system. With graded settings, and with the addition of directional relays, overcurrent relays can be made to work selectively on a loop system fed from one point. But on a loop fed sometimes from one point and sometimes from another, or on a network more complex than a loop, it is difficult, if not impossible, to choose settings for overcurrent relays so that the relays will work selectively for all fault locations and for all operating conditions.

New Words and Phrases

fuse　　　[fjuːz]　　　n. 熔丝
ascertain　[ˌæsəˈteɪn]　vt. 查明，弄清，确定
adjoin　　[əˈdʒɔɪn]　　vt. 贴近，毗连，毗邻
judicious　[dʒuˈdɪʃəs]　adj. 有见识的，明智的
versus　　[ˈvɜːsəs]　　prep. 对，对抗
solely　　[ˈsəʊllɪ]　　adv. 独自地，单独地
cascade　　[kæˈskeɪd]　　n. 串联，级联
breaker　　[ˈbreɪkə]　　n. 断路器
series trip coil　串联跳闸线圈
slow-speed overcurrent relay　延时过电流继电保护器
on account of　因为，考虑
circuit breaker　回路断路器
fault current　故障电流
transmission line　馈电线，输电线
time-current characteristic　时间-电流特性
graded time setting　阶梯形时间配制
line to line fault　相间故障

Exercises

Ⅰ. **Answer the following questions according to the text.**

1. What are the earliest methods of clearing faults?
2. What are the characteristics of the induction type of the overcurrent relays?
3. How does the relay operating time change when the fault is moved farther from the source of power?
4. How can the overcurrent relays be made to work selectively on a radial transmission or distribution system?

Ⅱ. **Translate the following phrases into Chinese.**

1. series trip coils　　　　　2. slow-speed overcurrent relays
3. on account of　　　　　4. circuit breaker
5. fault current　　　　　　6. time interval
7. transmission line　　　　8. time-current characteristic
9. graded time settings　　 10. line to line fault

Unit 9   Electric Power System Monitoring

Text

A modern electric power system is an assembly of many components, each of which influences the behavior of every other part. Proper functioning of the system as a whole makes it necessary to monitor conditions existing at many different points in the system in order to assure optimum operation.

The concern of the customers is primarily that the frequency and voltage of the supply are held within certain rather narrow limits. Since frequency of the system is the same everywhere, it may be monitored by a single frequency meter located at any convenient point. In contrast, the voltage of the system may be quite different at different points. Consequently, it is necessary to make continuous observation of the voltage at certain key points in the system in order to provide acceptable service.

Efficient operation of the system is obtained by assigning proper load schedules to each of the generators of the system. Newer plants, although individually more efficient, may be located at points of the system where their loading occasions large system losses. It is desirable to operate with such a division of the load between generators so that the total cost of fuel consumed is minimized. To provide reliability of the power supply in the event of unexpected conditions, it is desirable for all machines in operation to have the total kilowatt rating somewhat greater than the total load plus losses.

Instrumentation is necessary to permit billing of customers for energy used. Many interconnections exist between different power systems. Instruments must be provided at interchange points to permit billing for energy transferred from one system to another. The continuous monitoring of energy transfer is necessary to assure that interchanged power is within the limits of contract agreements.

The continuous measurement of conditions of major pieces of equipment is necessary to avoid damage due to overload. Instrumentation serves as a guide for future construction in a growing power system.

Occasionally, under emergency conditions, a system operator observes that his system load exceeds the ability of the available generating and transmission equipment. He is then faced with the problem of load shedding or, more properly, load conservation. It is then necessary to drop selected loads where service interruption is least objectionable. In such an event, he relies on the many instruments which provide information relative to system-operation conditions.

Instruments may sound alarms as advance warnings of conditions requiring action to avoid damage to equipment operating beyond its design limitation. In the event of extreme conditions such as power-system faults, defective equipment is switched out of service automatically. Instruments that continuously monitor current, voltage, and other quantities must be able to identify the faulted equipment and to bring about operation of the circuit breakers which remove it from service, while leaving in service all other equipment on the operating system.[1]

Many different electrical devices in a power system and those owned by the customers are designed for operation within certain specified ranges. Operation outside these designed limits is undesirable, as it may result in inefficiency of operation, excessive deterioration, or (in extreme cases) the destruction of the device. Careful attention to the conditions under which equipment is operating may indicate corrective action that must be taken.[2]

Overcurrent is undesirable for all electrical devices, as it produces excessive temperatures, inefficient operation and reduced service life. Overcurrent in residential circuits may bring about disconnection of the circuit by fuse or breaker action. Overcurrent in motors may also damage insulation.

New Words and Phrases

assembly   [əˈsemblɪ]   n. 装配组件，装配，安装
monitor   [ˈmɒnɪtə]   v. 持续观察，记录或测试（某物）的运作
optimum   [ˈɒptɪməm]   adj. 最佳的，最适宜的，最有利的
occasion   [əˈkeɪʒ(ə)n]   v. 致使，惹起，引起
division   [dɪˈvɪʒ(ə)n]   n. 分，分割，划分，歧异，差别，分化现象
minimize   [ˈmɪnɪmaɪz]   v. 使减（缩）小到最低，极力贬低，最低估计
instrumentation   [ɪnstrʊmenˈteɪʃ(ə)n]   n. 仪表化，测试设备
shed   [ʃed]   v. 去掉，除掉，脱落，剥落，蜕下
defective   [dɪˈfektɪv]   adj. 有缺点的，不完美的，不完全的
insulation   [ɪnsjʊˈleɪʃ(ə)n]   n. 绝缘
deterioration   [dɪˌtɪərɪəˈreɪʃn]   n. 恶化，降低，退化
result in   结果是，导致
contract agreement   合同协议书
frequency meter   频率计
transmission equipment   输电设备
power system fault   电力系统故障
circuit breaker   断路器
service interruption   停电

Notes

[1] Instruments that continuously monitor current, voltage, and other quantities must be able to identify the faulted equipment and to bring about operation of the circuit breakers which remove it from service, while leaving in service all other equipment on the operating system.

译文：用来监控电流、电压和其他量的测量仪器，必须能够辨别出产生故障的设备，使断路器开始起作用并将有故障的设备排除在供电服务之外，而让运行系统中其他所有设备继续运行。

说明：句中，that 引导一个定语从句，修饰主句主语 instruments，that 在从句中作主语。主句谓语是 must be able to identify... and to bring about..., 其中 the faulted equipment 和后面的 operation of the circuit breakers 分别是主句谓语复合结构中的两个宾语；which 在句中引导

另外一个定语从句，先行词是 the circuit breakers，which 在从句中作主语。

［2］Careful attention to the conditions under which equipment is operating may indicate corrective action that must be taken.

译文：认真观察设备运行状态才会知道应该如何采取正确措施。

说明：这是一个介词＋关系代词来引导的定语从句，句子语序原本应为 the conditions which equipment is operating under，但是由于 under 和 operating 不能构成固定的介词短语，因此必须把 under 提前，放在 which 前面。

Exercises

I. **Answer the following questions according to the text.**
1. Why do we have to monitor the electric power system?
2. What's the function of the instrumentation in the electric power system?
3. How can we monitor the frequency and the voltage of the electric power system?
4. What will the overcurrent cause for all the electrical devices and the residential circuits?

II. **Decide whether the following statements are true（T）or False（F）according to the text.**
1. We can monitor the voltage of the power system at any convenient point.
2. Overcurrent can produce excessive temperatures, inefficient operation, and reduced service life for all electrical devices.
3. To provide reliability of the power supply in the event of unexpected conditions, it is desirable for all machines in operation to have the total load plus losses somewhat greater than the total kilowatt rating.
4. In order to avoid damage due to overload, we have to monitor the conditions of major pieces of equipment occasionally.
5. The many different electrical devices in a power system and those owned by the customers are designed for operation within certain specified ranges. Operation outside these designed limits is undesirable.

III. **Translate the following expressions into English.**
1. 频率计   2. 输电设备
3. 电力系统故障   4. 断路器
5. 停电   6. 系统负载
7. 限电；减载；甩负荷 8. 电气设备

IV. **Translate the following sentences into English.**
1. 汽车速度不得超过每小时 55 英里的最大时速。
2. 企业必须密切关注社会的变化和需求，以便更好地承担社会责任。
3. 我们要应对这些挑战，最重要的是靠发展。
4. 注意力要集中在获得所需答案应做的工作上。

Translating Skills

被动语态的译法

科技英语重在描述客观事实,阐述科学技术的成果、实验研究的过程和数据,所以科技英语在句子结构方面有其独特之处,最明显的特点就是被动语态的大量运用。被动语态常见的翻译方法如下:

1. 译成汉语主动句

1)用"人们""我们"等含有泛指意义的词做主语,从而使汉语译文呈"兼语式"句式。

The mechanism of fever production is not completely understood. 人们还不完全清楚发烧的产生机理。

He was seen to have turned off the current. 有人看见他把电流断开了。

2)由 it 引导的主语从句,译成无人称或不定人称句。

It is said that the production of transistor radios was increased six times from 1970 to 1974. 据说1970~1974年间晶体管收音机的产量增长了5倍。

3)当被动语态中的主语为无生命的名词或由 under、in、on 等少数几个介词短语构成时,译成主动句。

That computer is under repair. 那台计算机正在修理。

4)当 need、want、require 等动词后接主动形式的动名词,表示被动意义时,译成主动句。

This device needs repairing. 这套设备需要修理。

2. 译成汉语被动句

1)译成明显的"被"字句。

In this case, the molecule is polarized by the field. 在这种情况下,该分子被场极化了。

2)译成"由……组成、用于……"等,通常带有"be made up of""be composed of""be used to"。

The magnetic field is produced by an electric current. 磁场由电流产生。

3)译成判断句式"是……的"。这种判断句能清楚表达作者的意图和客观事实。

Everything in the world is built up from atoms. 世间万物都是由原子构成的。

4)在谓语动词前面添加"予以""加以""受到""得以"等词。

Coal and oil are the remains of plants and animals. Crude mineral ores and crude oil must be purified before they can be used. 煤和石油是动植物的残骸。原矿石和原油必须加以精炼才能使用。

3. 译成汉语无主句

Work is done, when an object is lifted. 当举起一个物体时,就做了功。

Attention must be paid to safety in handling radio active materials. 处理放射性材料时必须注意安全。

在翻译被动语态语句时,决不能固守原句,要灵活地采用多种翻译技巧,运用各种修辞手段,能译成主动句的就译成主动句,须译成被动句的就译成被动句,根据汉语的语法和习惯，用规范化的汉语表达方式忠实而恰当地反映出原作的真实含义，使译文的形式与原文内容辩证地统一起来，才能收到良好的翻译效果，使译文既在内容上忠实于原文，又符合汉语的表达方式。

Reading

Distribution Automation Increases Reliability

The Long Island Lighting Co. (LILCO) provides service to over a million customers via 750 plus distribution feeders, comprised, mainly, of overhead lines. These feeders are subject to the usual storm-related problems involving damage due to thunderstorms, lightning, tree contact, ice storms and hurricanes. Compared with other New York State electric utilities, on average, LILCO provides the fastest restoration, although customers have in the past experienced more frequent sustained interruptions.

An analysis of customer outages determined that about 78% of interruptions were due to faults on main three-phase lines. A fault on the main line would cause, on average, 2000 customers to lose power. In order to reduce these outages, the company has instituted reliability programmes that include trimming trees, installing lightning arresters, installing covered wire and replacing obsolete armless insulators. These programmes reduced outages, but were not sufficient to reach the desired level of service continuity. To further improve reliability, LILCO has installed an advanced distribution automation system, developed jointly by LILCO and Harris Distributed Automation Products, Calgary, Alberta, Canada. This system isolates faults and restores non-damaged portions of the main line circuit based on real-time parameters of voltage, current, breaker status and supervisor-controlled switch positions. Since this project was initiated in 1993, more than 240,000 customers have avoided a sustained interruption of service.

In the early 1970s, LILCO installed over 400 pole-mounted switches with radio controls. This equipment which included a mix of medium voltage vacuum switches and current sensors reduced outage duration but had little impact on reducing outage frequency. The autosectionalizing phase required the development of a remote terminal unit (RTU) with local intelligence that could detect permanent load-side faults and open a motor-operated switch prior to the substation breaker locking out. The Harris DART RTU was selected for this operation, providing transducerless AC inputs that would directly connect to the existing sensors. In addition, the DART has a patented fault detection algorithm, which was developed and tested by S & C Electric CO., Chicago, Illinois, U.S. This fault detection scheme can detect three levels of overcurrent, including overcurrent with no loss of voltage, overcurrent with a successful breaker trip reclose and overcurrent with a breaker lockout.

Building upon the fault detection scheme, Harris developed an advanced RTU application, known as the "Smart Switch" or "Autosectionalizer." This advanced application requires no communications network or relay coordination study to be implemented. Integration of the Autosectionalizers to the 400 existing switches was completed in April of 1996. In addition, LILCO has installed over 350 S & C SCADAmate switches with Autosectionalizer capability. These 750 Autosectonalizing switches enable LILCO to reduce by 25% the number of customers affected by main line faults, amounting to 240,000 customers avoiding a sustained outage in the last 18 months.

The second stage of the project involved the installation of a Harris PC-based SCADA system along with setting up communications linked with the Autosectionalizing switches, which gave the operators remote indication and control of switches. The Harris PC SCADA provides, on a three-phase basis, voltage, current, power factor, load profiles and other real-time values for optimizing the use of the distribution system. After evaluating several communication mediums, LILCO selected Microwave Data System (MDS) spread spectrum radio network. The MDS radio incorporates unlicensed frequency hopping spread spectrum technology. Using the MDS system, LILCO installed 750 radios and established 90% communications coverage in just under nine months.

The third stage of the project required the development of an advanced algorithm for the automatic restoration of power reaching the non-faulted feeder zones. This algorithm was written to run on the Harris D-200, which serves as the front end processor for communications with the RTUs.

The Harris D-200 receives information from the Autosectionalizing switches reporting loss of voltage and fault location. In addition, the Harris D-200 receives substation breaker status and loading from the affected circuit as well as information from the adjacent tie circuit. Finally, based on real-time voltage, load, breaker status, switch positions and safety interlocks, the algorithm calculates and reports corrective measures to restore service. The autorestoration algorithm is designed to restore up to 12 feeders simultaneously and will support as many as seven switches arranged in an open loop arrangement. In addition, the operators can set the level of automation for each circuit loop. The levels include:

1) The manual. The PC SCADA system will not issue a recommendation to restore service. The operators have full supervisory control and indication of switches.

2) Operator Acknowledgement. The Harris D-200 will determine the faulted zone and then based on real-time information will issue to the PC SCADA a simulation screen and the switching steps required to restore service. If the operator agrees with the algorithm's recommendation, the operator "issues an acknowledgment" enabling the PC SCADA to perform the necessary switching steps. A failure of any switching step aborts the restoration and places the circuit loop into manual.

3) Full Automation. Based on the algorithm the faulted zone is determined and the switching is automatically performed to restore the non-faulted zones. All automatic switching steps are based on real-time information, and operator procedures (interlocks) prevent the energizing of work areas. No operator intervention is required.

On April 30, 1996, LILCO received a patent for the advanced restoration algorithm. While many utilities are using supervisory switching to isolate faults, most use operator remote control to restore service. This procedure is time consuming and inefficient, particularly during storm conditions when multiple faulted circuits occur.

New Words and Phrases

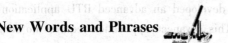

restoration   [ˌrestəˈreɪʃ(ə)n]   n.   回复到原处或原状，恢复，修复，整修
sustained   [səˈsteɪnd]   adj.   持久的，经久不衰的
outage   [ˈaʊtɪdʒ]   n.   停机，断电

arrester ［əˈrestə］ n. 逮捕者，制动器，避雷装置
obsolete ［ˈɔbsəli:t］ adj. 不再使用的，过时的
parameter ［pəˈræmitə］ n. 参量，参数
initiate ［ɪˈnɪʃɪeɪt］ v. 创立，引进
transducer ［trænzˈdju:sə］ n. 转换器，传感器
patent ［ˈpeɪt(ə)nt］ adj. 有专利的，受专利权保护的
algorithm ［ˈælɡərɪð(ə)m］ n. 规则系统演段
spectrum ［ˈspektrəm］ n. 光谱，频谱，范围，幅度
adjacent ［əˈdʒeɪs(ə)nt］ adj. 毗连的，接近的，相接触的
abort ［əˈbɔ:t］ v. 中止，夭折
be subject to 遭受
obsolete armless insulator 过期无防护的绝缘器
real-time parameter 实时参数
a permanent load-side fault 永久负载侧故障
RTU（Remote Terminal Unit） 远程终端设备
SCADA（Supervisory Control And Data Acquisition） 监控数据探测
to reach the desired level of service continuity 达到理想的持续供电标准
spectrum radio network 波谱无线电网络
auto-restoration algorithm 主动恢复供电算法
non-faulted feeder zone 无故障供电范围
auto-sectionalizing switches 主动分段开关
substation breaker status 变电站断路器状况
the adjacent tie circuit 临近线路
full supervisory control 完全监控
algorithm's recommendation 算法推荐（操作）
MDS（Microwave Data System） 微波数据系统

Exercises

Ⅰ. **Answer the following questions according to the text.**

1. What causes the service interruption in most cases?

2. What's the function of the RTU application, known as the "Smart Switch" developed by Harris?

3. How many levels of automation can we set for each circuit loop when Harris D-200 is used? And what are they?

Ⅱ. **Translate the following phrases into Chinese.**

1. lightning arrester
2. a sustained interruption of service
3. a permanent load-side fault
4. multiple faulted circuits
5. full supervisory control
6. circuit loop
7. auto-sectionalizing switches
8. RTU（Remote Terminal Unit）
9. real-time parameter
10. a three-phase basis

Chapter II   Intelligent Building Technology

Unit 10   Intelligent Building

Text

   Intelligent building successfully merges building management and IT systems to optimize system performance and simplify facility operations, it designs and offers a safe, comfortable, energy-conservation, high-efficient, convenient working environment, go on after the analysis that optimize most through structure, system, service, management in building, etc.. It integrates traditional architecture and modern 4C technology (Computer, Control, Communication and CRT graphical display technology). It has already become the development mainstream of the construction in the 21st century.

   The intelligent building is made up of Building Automation System, Communication Automation System, Office Automation System, Structured Cabling System, System Integration, Security Automation, Fire Control Automation. The purpose is to reach safety, comfort, energy-conservation, high-efficient, environmental protection.

   Building Automation Systems (BAS) use computer-based monitoring to coordination, organize and optimize building control sub-systems such as security, fire/life safety, elevators, etc. It is mainly composed of HVAC control systems, building water supply and drainage control systems, building control systems for power distribution, lighting control systems, traffic control systems, fire prevention and security systems, construction equipment automation systems integration. Applications are as follows:

   1) Equipment scheduling (turning equipment off and on as required);

   2) Optimum start/stop (turning heating and cooling equipment on in intelligence to ensure the building is at the required temperature during occupancy);

   3) Operator adjustment (accessing operator set-points that tune system to changing conditions);

   4) Monitoring (logging of temperature, energy use, equipment start times, etc.);

   5) Alarm reporting (notifying the operator of failed equipment, out of limit temperature/ pressure conditions or need for maintenance).[1]

   Communication network system includes communication network and information network system. Communication network is the infrastructure of voice, data, and video transmission in a building, it is possible that connects external communication network (such as the public telephone network, integrated services digital network, Internet, data communication networks and satellite communication networks, etc.) to achieve open information and information sharing through the communication networks. In formation Network System is the information network platform consisting of advanced technology and equipment,[2] which apply computer technology, communication technology, multimedia technology, information security technology, behavioral science and advanced technology and equipment.

Office Automation System is composed of multi-functional telephones, high-performance fax machines, various types of terminals, PCs, word processors, the main computers, audio-visual storage devices. Raw data storage, electronic transfer, and the management of electronic business information are the basic tasks of an office automation system, which is used to digitally create, collect, store, manipulate, and relay office information needed for improving a working environment.[3]

System Integration Center regards all the elements of intelligent building as a core element. It uses the integrated structured cabling system and computer network technology to design synthetically language, image and data signals in a set of integrated structured cabling system through unified planning, and it uses the premises distribution systems and public communications network inside and outside the building as a bridge, as well as the coordination of various system and the interface between the LAN and agreements, turning the separation of the equipment, features and information into a number of organic connection as a whole, thus, constituting a complete system for a high degree of resource sharing and focus.

Structured Cabling System is constituted with variety of components as an organic whole, which takes the modular design of the structure and supports the transmission of voice, data and image, thus, it meets the requirements for information transmission. SCS is composed of six sub-systems: the sub systems of work area, horizontal cabling, management, backbone cabling, campus, and equipment room.

The purpose of intelligent building is with modern 4C technology to form structure and system of intelligent building, combine the modernized service and management style, and provide a safe, comfortable life, study and working environment for people.

New Words and Phrases

intelligent [ɪnˈtelɪdʒ(ə)nt] adj. 聪明的，理解力强的，智能的
merge [mɜːdʒ] vt. & vi. （使）混合，（使）合并
optimize [ˈɒptɪmaɪz] vt. 使最优化，使完善，使尽可能有效 vi. 优化，持乐观态度
facility [fəˈsɪləti] n. 设备，便利的设施
conservation [ˌkɒnsəˈveɪʃ(ə)n] n. 保存，保持，保护
architecture [ˈɑːkɪtektʃə] n. 建筑学，建筑式样，建筑风格
coordination [kəʊˌɔːdɪˈneɪʃən] n. 对等，同等，协调，调和
distribution [ˌdɪstrɪbjuːʃ(ə)n] n. 分配，分布
optimum [ˈɒptɪməm] adj. 最适宜的，n. 最佳效果，最适宜条件
infrastructure [ˈɪnfrəstrʌktʃə] n. 基础设施，基础结构
manipulate [məˈnɪpjʊleɪt] vt. 操作，操纵，巧妙地处理，篡改
relay [ˈriːleɪ] v. 转播，接替，n. 继电器，接替，接替人员
premise [ˈpremɪs] n. 处所，楼宇，前提
interface [ˈɪntəfeɪs] n. 接口，界面

organic	[ɔːˈgænɪk]	adj 有机的，器官的
constitute	[ˈkɒnstɪtjuːt]	vt. 组成，构成，任命，建立
modular	[ˈmɒdjʊlə]	adj. 模的，模块化的，有标准组件的
horizontal	[ˌhɒrɪˈzɒnt(ə)l]	adj. 水平的，地平线的
backbone	[ˈbækbəʊn]	n. 决心，毅力，支柱，脊椎，主干网
be constituted with		由……组成

Technical Terms

Communication Automation System	通信自动化系统
Structured Cabling System	综合布线系统
HVAC	暖通空调系统
System Integration Center	系统集成中心
Integrated Services Digital Network	综合业务数字网

Notes

[1] ... notifying the operator of failed equipment, out of limit temperature/pressure conditions or need for maintenance.

译文：通知操作员设备故障、温度/压力超标或者需要维护。

说明：句中 notify sb. of sth. 正式将（某事）通知（某人或某团体）。out of limit 失去控制。又如：

I have notified the police of a loss. 我已向警察报告了损失情况。

[2] Information Network System is the information network platform consisting of advanced technology and equipment.

译文：信息网络系统是应用先进技术和设备构成的信息网络平台。

说明：句中动名词短语 consisting of advanced technology and equipment 作后置定语，修饰 platform。

[3] ... which is used to digitally create, collect, store, manipulate, and relay office information needed for improving a working environment.

译文：办公自动化系统通过数字化创建、收集、储存、处理和传输办公所需的资料，改进工作环境。

说明：句中 needed for improving a working environment 是过去分词短语作后置定语，修饰 information。

Exercises

I. Choose the best answer.

1. (　　) has already become the mainstream development of the construction in the 21st century.
   A. Large building　　　　　　　　B. Fashion building
   C. Intelligent building　　　　　　D. Complex building

2. Building Automation System based on (　　) to coordinate, organize and optimize building

control sub-systems.

  A. information   B. computer   C. operator   D. rules

 3. ( ) are the basic tasks of an office automation system.

  A. Raw data storage   B. Electronic transfer

  C. Management of electronic business information   D. All of the above

 4. Structured Cabling System is constituted with ( ) as an organic whole.

  A. different components   B. computer systems

  C. lighting control systems   D. fire prevention and security systems

 5. The modern 4C technology refers to ( )

  A. computer and control   B. communication and CRT

  C. computer and communication   D. A and B

Ⅱ. Decide whether the following stataements are true (T) or False (F) according to the text.

 1. It is a product of traditional architecture and modern 4C technology.

 2. Communication Network System includes information system network and office automation.

 3. System Integration Center regards the main computers as a core element.

 4. PDS is composed of six sub-systems. They are systems of work area, horizontal cabling, management, backbone cabling, campus, and equipment room.

 5. Information network is the infrastructure of voice, data, and video transmission in a building.

Ⅲ. Translate the following into English or Chinese.

 1. 楼宇自动化系统   2. 暖通空调系统

 3. 采取模块化结构设计   4. 综合业务数字网

 5. System Integration Center regards all the elements of intelligent building as a core element.

 6. Intelligent building has already become the development mainstream of the construction in the 21st century.

Translating Skills

非谓语动词 V-ing 的用法

  在英语中，一句话中当要叙述几个动作时，先选其中一个主要的动作作为谓语，用动词表示，其余动作要用动词的变形，称为非谓语动词。非谓语动词通常有 3 种：V-ing，V-ed 和 to V。

  V-ing 形式的动词在句中可以作状语、主语、宾语和定语，同时保留了动词性，因此可带有宾语和状语。

1. 作定语

  单个 V-ing 作定语一般放在名词前面（也可以放在后面），V-ing 短语作定语一般放在名词之后。且 V-ing 本身含有主动进行的意思，表示动作是由所修饰的名词主动发出的。

  Devices <u>utilizing</u> a static voltage as the <u>controlling</u> signal are called voltage-controlled devices. 利用一个电压作为控制信号的器件叫作电压控制器件。

  这里 V-ing 短语作定语，utilizing（利用）的动作是由器件发出的。

  A diode is an electrical device <u>allowing</u> current to move through it in one direction with greater ease than in the other. 二极管是一种电流正向流通比反向流通要容易得多的器件。

这里 V-ing 短语作定语，allowing 是由 device 发出的动作，根据专业知识，也可以译成：二极管具有电流单向导通的特点。

2. 作状语

V-ing 短语作状语时，往往具有时间、条件、原因、结果、方式、补充说明等含义，可放在句首、句中或句尾，通常它的逻辑主体就是句中的主语。

When placed in a simple battery-lamp circuit, the diode will either allow or prevent current through the lamp, <u>depending</u> on the polarity of the applied voltage. 当在一个简单的电池—电灯电路中串联一个二极管后，根据所加的电压的极性不同，二极管会出现导通电流或阻碍电流流过电灯的现象。(V-ing 短语作原因状语)。

分词短语作状语时，前面还可用 when、while、if、unless、though 等连词来加强时间、条件等含义。

When <u>measuring</u> current, the circuit must be opened and the meter inserted in series with the circuit or component to be measured. 当测量电流时，必须断开电路，将万用表与待测电路或元器件串联。

3. 作主语或宾语

<u>Heating</u> the water changes it into vapor. 把水加热可以使水变为蒸汽。(V-ing 短语作主语)

It may also have a polarity switch to facilitate <u>reversing</u> the test leads. 还有一个极性开关可以很方便交换测试笔的极性。(V-ing 短语作 facilitate 的宾语)

4. 作主语或宾语的补足语

We put a hand above an electric fire and feel the hot air <u>rising</u>. 我们把手放在电炉的上方，就会感觉到热空气在上升。

5. 与 with（without）连用

在科技文章中，常用"with（without）+名词+分词"结构用做补充说明。

The density of air varies directly as pressure, with temperature <u>being</u> constant.

在温度不变时，空气密度与压力成正比。

Reading

Intelligent Hotel

The intelligent design of advanced hotels focuses on the high starting point, block time, high dependability, leaving the abundant expansion space, for constructing of the intelligent hotel, can offer more overall service and faster response speed of service to households and guests who move in hotels, and makes the living environment of guests and households safer, comfortable, convenient, high-efficient, energy-conservation, flexibility, environmental protection. Thus it improves the service level of hotels and management level greatly and can utilize intelligent systematic energy-conservation, reduce and operate the expenses, raise the management level of hotels and economic benefits.

Brief introduction of subsystem of "Intelligent system of hotels" is as follows:

1. Hotel's VOD information service network system of multimedia

VOD technology is a technology which let people request and interchange the information of the multimedia as required whenever and wherever possible. Setting up high speed interactive VOD network system of broadband is a supreme goal of the social development of the network, and is the condition of realizing the ideal of global village. Hotel's VOD information service network system of multimedia is the guests that use for inquiring and booking hotels to serve, browsing through all kinds of online information, viewing and admiring the film, going to Internet. This system can be fully realized "people first", "personalized service", "service with information", "convenient, comfortable" these service theories in information age of network.

2. Building automatic control system of the hotel

This system is used for automatic control of the environmental parameter of the hotel guest room and public place, for instance: temperature, humidity, new trend, smell, except that fungus, etc.. The purpose is to create a comfortable, warm accommodation environment for guests with home from home feelings.

3. Administrative system of the guest room

This system is mainly used in illumination, stereo of the room, the TV controller, serve requests, no disturb, etc..

4. Commercial comprehensive administrative computer system

This system is mainly used in hotel reservation and management and administration of the chain, front desk, backstage supporter's computer management, scheduled, check out, financial affairs, report form, inquiry, data base administration, EMAIL mail management.

5. All-purpose card of hotels, entrance guard's system

This system is used mainly in:
1) Guest's identity discerning
2) Guest's consuming and keeping accounts management
3) The management of the historical record consumed by guest and his discount
4) The management of the guest's personalized service
5) The management of hotel security, lock controlling

Because of the adoption of this system, it can accomplish humanization, more individualized service and management for guests. Senior guests can use the non-contact-type radio frequency card, which makes guests enjoy tight following and safeguarding unconsciously. The system controls the VIP district and the person without radio frequency card is unable to take action at will after entering.

6. Hotel deal in and office automated system

This system is set up on the WAN/LAN, offering high-quality management to the hotel administrator, such as intellectual office system, intellectual energy-conservation system, intelligence purchase network, intellectual administrative systems of personnel, intellectual material consumption administrative system, etc. The purpose is to make hotel official work, material consumption, energy consumption, personnel's cost lowest. It is most effective to use and create well benefit.

7. Structurization wiring system

This system is the information passway, on which all the "intellectual weak electricity system of hotels" based on WAN/LAN are operated. It is the expressway of the network system, and is a neural line of the weak electricity system.

8. **Communication system**

This system is used in:
1) Guest's communication to the external
2) Communication inside hotels

The good communication network can make guests carry on the information transmission of the multimedia, such as language, picture, data, etc. It can hold meeting of the network, making the videophone, accessing the Internet, etc. It makes guests feel in an open, convenient information intensive society. Even travelling outside, they have the same feeling at home.

New Words and Phrases

supreme   [suː'priːm]   adj. 最高的，至上的，最重要的
parameter   [pə'ræmɪtə]   n. 因素，特性，界限，参量，参数
fungus   ['fʌŋɡəs]   n. 真菌
administrative [əd'mɪnɪstrətɪv] adj. 行政的，管理的
illumination [ɪˌljuːmɪ'neɪʃən] n. 照明，强度，彩灯，灯饰
expressway   [ɪk'spreswei]   n. 高速公路
neural   ['njʊər(ə)l]   adj. 神经的

Exercises

Answer the following questions according to the text.

1. What is the purpose of the intelligent hotel?
2. What's the function of the hotel's VOD information service network system of multimedia?
3. Where is the automatic control system used for?
4. What's the function of the office automated system?
5. How about the feeling of a guest living in an intelligent hotel?

Unit 11 Structured Cabling System

Text

Structured Cabling System (SCS) is an absolutely necessarily subsystem of the intelligent building. It is the system to transfer the information of voice, data and image. A well-designed cabling system should be open, flexible, expansible, and independent from the equipment. The system meets the rapid development of the technology and ensures the customers to gain the reliable guarantee of their investment.

Structural integrated cabling system has become the basic facilities of a building nowadays. It is the medium of information transmission in a building or among buildings. It connects not only the communication equipment, exchange equipment of voices and data, and other information management system, but also other sub-systems of intelligent building. Its flexibility, compatibility and reliability are recognized by customers in China. It has already been applied in the government departments, banks, large-scale enterprises, real estate industry, etc.. Structural integrated cabling provides customers ideal way of cabling, relying on its high quality material and innovative cabling methods,[1] which lays a solid foundation for the modern buildings and factories to become the real intelligent type buildings, whose transmission medium of communication stays unchanged for 15 years.

SCS consists of six sub-systems:

1) Work area sub-system is made of cabling from the terminal devices to the information outlet. The terminal devices include computers (PC, workstation, server, printer etc.), telephones, fax machines, duplicating machines and so on.

2) Administration sub-system is made of the connected patch panels, which facilitates the system management for the administrator to realize cabling management. It is easy to track the patch cord due to its perfect design, saving 50% of space for the cabinets at each floor or patch panel, compared with the traditional patching case owing to its small size.[2]

3) Horizontal sub-system adopts the Cat. 5e or Cat. 6e cable to connect the information outlet and administration sub-system.

4) Backbone sub-system connects administration sub-systems from equipment management center to each floor. It is suggested to adopt the fiber cable as the backbone to connect the core administrative sub-system to the equipment center. As for the backup backbone, Cat. 6e copper cables are recommended, which enable that the system transfer speed reaches as high as 1G B/s.[3]

5) Equipment sub-system connects the network cabling systems and primary equipment, by outfitting relative adapters for different equipment which include the Ethernet Switch, PABX, Router, server and firewall configured in the equipment center.

6) Building sub-system connects the buildings.

The cost of an integrated network cabling system is only 10% of the whole network construction; however, nearly 70% of network problems come from the inferior cable parts and cabling technology. So, a well-designed and well-organized Structured Cabling System can save a great deal of capital and manpower during the process of installation, maintenance and upgrading, with its total cost of ownership decreased by 30% on average. As a result, a well-integrated network cabling system will facilitate our work and life.

New Words and Phrases

cabling ['keɪblɪŋ] n. 卷缆柱,卷绳状雕饰
compatibility [kəmˌpætɪ'bɪlɪti] n. 兼容性
estate [ɪs'teɪt] n. 土地,地区,庄园,种植园,地产,财产,遗产
innovative ['ɪnəvətɪv] adj. 新发明的,新引进的,革新的
duplicate ['djuːplɪkeɪt] vt. 复制,复印 adj. 完全一样的
panel ['pænəl] n. 面,板,控制板,仪表盘
facilitate [fə'sɪlɪteɪt] vt. 使便利,减轻……的困难
cord [kɔːd] n. (细)绳,灯芯绒裤
cabinet ['kæbɪnɪt] n. 橱,陈列柜,内阁,内阁会议
adopt [ə'dɒpt] vt. 收养,采用,采纳,采取,正式接受,通过
cable ['keɪb(ə)l] n. (船只、桥梁等上的)巨缆,钢索,电缆
adapter [ə'dæptə] n. 适配器,改编者
configure [kən'fɪgə] v. 配置,设定,使成形,使具一定形式
inferior [ɪn'fɪərɪə] adj. 低等的,下级的,劣等的,次的 n. 部下,下属
maintenance ['meɪnt(ə)nəns] n. 维持,维护,保养,维修,赡养费
upgrade [ʌp'greɪd] n. 向上的斜坡,vt. 提升,使升级
installation [ˌɪnstə'leɪʃ(ə)n] n. 安装,装置,设备,军事设施

Technical Terms

structured cabling system	结构化布线系统
real estate	房地产;房地产所有权
work area sub-system	工作区子系统
information outlet	信息插座
administration sub-system	管理子系统
patching case	配线箱
B/s (Bits per second)	位/秒
patch cord	跳线
PABX	程控交换机
Ethernet Switch	以太网交换机

Notes

[1] Structural integrated cabling provides customers with ideal way of cabling, relying on its high quality material and innovative cabling methods....

译文：结构化综合布线系统为用户提供了理想的布线方式，并依靠其高品质的材料，一改传统的布线方法……

说明：句中 relying on its high quality material 分词短语作条件状语。rely on 译为"依靠，依赖"，又如：

This ensures that customers can rely on the rigid quality standards and that the company can meet the high quality expectations of the AHT brand. 这保证了客户可以信赖我们严格的质量标准，并且我们公司也可以满足 AHT 这一品牌对质量的高要求。

［2］ It is easy to track the patch cord due to its perfect design, saving 50% of space for the cabinets at each floor or patch panel, compared with the traditional patching case owing to its small size.

译文：它很容易追踪跳线，体积小，比传统配线箱节省 50% 空间。

说明：句中 it 作形式主语，真正的主语是 to track the patch cord。过去分词短语 compared with the traditional patching case 在句子中作比较状语。

［3］ As for the backup backbone, Cat. 6e copper cables are recommended, which enable that the system transfer speed reaches as high as 1GB/s.

译文：至于备用主干线，可以采用超六类双绞铜线，其系统传输率可高达 1GB/s。

说明：句中 which 引导的是非限制性定语从句，which 指代前面的 Cat. 6e copper cables。bit/s 每秒传递位数（bit per second）。

Exercises

Ⅰ. **Answer the following questions.**

1. What's the role of SCS?
2. Where has the SCS been applied?
3. What are the sub-systems of SCS?
4. Which sub-system can connect the information outlet and administration sub-system?
5. What's the role of building sub-system?

Ⅱ. **Translate the following phrases into English.**

1. 结构化综合布线系统　　2. 交换设备　　3. 信息传输
4. 终端设备　　　　　　　5. 信息管理系统　6. 信息插座
7. 网络布线系统　　　　　8. 以太网交换机

Ⅲ. **Translate the following sentences into Chinese.**

1. It is the system to transfer the information of voice, data and image.
2. Work area sub-system is made of cabling from the terminal devices to the information outlet.
3. It is easy to track the patch cord due to its perfect design, saving 50% of space for the cabinets at each floor or patch panel, compared with the traditional patching case owing to its small size.
4. As a result, a well-integrated network cabling system will facilitate our work and life.

Translating Skills

非谓语动词 V-ed 和 to V 的用法

1. V-ed 形式

动词的 V-ed 形式与 be 结合构成被动语态，与 have（had）结合构成完成时，因此 V-ed 形式本身含有被动与完成的意思，它可在句子中担任定语、表语、状语等，保留了动词性，表示这一动作是已完成的或所修饰名词所（被动）接受的。

1) V-ed 作定语。

单个 V-ed 作定语一般放在名词前面（也可以放在后面），V-ed 短语作定语一般放在名词之后。在作定语时，分词在意思上接近一个定语从句。

The skills <u>acquired</u> are useful in electrical engineering areas. 获得的技能在电气工程领域是十分有用的。

A semiconductor diode consists of a PN junction <u>made of semiconductor material</u>. 一个半导体二极管是由一个半导体材料制成的 PN 结构成的。

2) V-ed 作表语。

The students will get <u>confused</u> if they are made to learn too much. 如果让学生学得太多，他们会感到糊涂的。

3) V-ed 作状语，表示时间、原因、条件、方式或动作发生的背景和情况。

<u>Observed</u> from the spaceship, our earth looks like a blue ball. 从宇宙飞船上看，我们的地球像一个蓝色的球体。

<u>Compared</u> to a conventional process control system, number of wires <u>needed for connections</u> is reduced by 80%. 与传统的过程控制系统比较，所需的连接线减少了 80%。

2. to V 形式

To V 又称为动词不定式，兼有名词、形容词、副词的特点，也保留了动词性，可用作主语、宾语、表语、定语、状语，常用来表示具体的（特别是未来的）一次性动作。

Resistor is made variable to be able to adjust sufficient feedback voltage to cause oscillation. 电阻是可调的，可调节到有足够的反馈电压以引起电路的振荡。

这里 is made 不译出，"电路的"为增补词语。用 to V 表示一次性的、未来的动作。

It is very difficult <u>to measure the passing current in insulators</u>. 测量绝缘体中通过的电流是很困难的。

动词不定式短语作主语时，尤其是动词不定式短语比较长时，往往引入 it 作形式主语，而把动词不定式短语放在谓语动词的后面。

<u>To do this</u>, we must introduce the concept of a "loop". 为了这么做，必须引入"回路"的概念。

这里动词不定式短语作状语。

Tesla's work on induction motors and poly-phase systems influenced the field for years <u>to come</u>. 特斯拉关于感应电动机和多相输电系统的研究到今天还影响着电气工程领域。（这里 to V 作定语，修饰 years，意为"未来的年代"。）

在翻译中要特别注意 to V 与 V-ing 的区别，否则会造成误解。如：

Stop to smoke. 停下来抽一支烟（一次性动作）。
Stop smoking. 戒烟（终止这一经常性的动作）。
I forgot to do it. 我忘记做这件事了（事没有做）。
I forgot doing it. 我忘记做过这件事了（事已做了）。

Reading

Solutions of Home Structured Cabling System

Pre-wiring an intelligence home is no longer a luxury. It is a necessity. PUTIAN Home Structured Cabling System is the foundation on which your home communications and entertainment network will be built. It is designed to organize and distribute your new home network. Structured module connects each room on Cat. 5e or Cat. 6e cable to a central spot called PUTIAN Distribution Box. This facility then manages and distributes voice, data, audio, and video also security signals throughout your home.

PUTIAN Home Structured Cabling System has unified management of the intelligent home equipment signals, such as the net, telephone, cable television, video, monitor and so on (includes the outlet, cables and Family Multimedia Distribution Box). This system is the passage of the information of every equipment inside the intelligent home. It is the basis of realizing the different functions.

PUTIAN Home Structured Cabling System is equal to the family vision from comprehensive cabling system's working area to the management area. It also includes the working area, transport area and the management area. It is one part of the comprehensive structured cabling system. Their difference is the more types of the information station. Besides the language information station, data information station, it also has the cable television information station, home theater (background music) information station, security or the monitor information station.

The standards of the home wiring provide a telecommunication service to meet the minimum requirements of the common cabling system, and the grading can provide telephone, cable television and data services. A star rating is in accordance with the topology, and is used of non-unshielded twisted pair connections. Here the use of non-unshielded twisted pair EIA/TIA-568A must meet or exceed the provisions of the three types of cable transmission requirements. In addition, it will be a 75-ohm coaxial cable, and must meet or exceed the SCTE IPS-SP-001's request to transmit television signals. It is proposed to install ultra-category 5 unshielded twisted pair (UTP) for upgrading grade 2 in the future. Specifically configured to:

1) Per household can introduce a super-five pairs of twisted pair cables; Synchronous (makes residential customers within the neat appearance, for the construction and maintenance, user-friendly) the laying of a 75-ohm coaxial cable and the corresponding socket, introduce the Family Information Distribution Box.

2) Each household to set up information embedded niche family wiring box.

3) Each bedroom, study, sitting room, restaurants should set up an information outlet (for data or voice communications) and a cable television outlet, main bathroom and other rooms should be

set up for telephone information outlets.

4) Each information or cable television outlet to outlet and Family Information Distribution Box, the laying of a Category 5 four pairs of twisted pair cables or a 75-ohm coaxial cable, set up a satellite-based network.

5) If the security system access to Family Information Distribution Box, in accordance with the requirements of the various security system corresponding laying of cables and configure the corresponding terminal equipment.

6) Family Information Distribution Box configuration should be a place to meet long-term needs.

New Words and Phrases

luxury ['lʌkʃ(ə)rɪ] n. 奢侈，豪华，奢侈品
unify ['juːnɪfaɪ] vt. 使联合，统一，使相同，使一致
comprehensive [kɒmprɪ'hensɪv] adj. 广泛的，综合的
rating ['reɪtɪŋ] n.（船上人员的）等级，类别，（海军）水兵
topology [tə'pɒlədʒɪ] n. 拓扑，布局，拓扑学
twist [twɪst] vt. & vi. 扭，搓，缠绕 vt. 转动，拧，歪曲，曲解
provision [prə'vɪʒ(ə)n] n. 供应，提供，供给，规定，条款，条件
in accordance with 依照，根据
structured module 布线模块
distribution box 配线箱
data information station 数据信息点

Exercises

I. **Answer the following questions according to the text.**
1. What is the basis of realizing the different functions?
2. What kind of information stations are in the PUTIAN Home Structured Cabling System?
3. What makes up of the management area?

II. **Translate the following Sentences into Chinese.**
1. PUTIAN Home Structured Cabling System is the foundation on which your home communications and entertainment network will be built.
2. PUTIAN Home Structured Cabling System is equal to the family version from comprehensive cabling system's working area to the management area.
3. A star rating is in accordance with the topology, and is used of non-unshielded twisted pair connections.

Unit 12　Office Automation System

Text

Office Automation System is composed of correspondence management, document-sending management, document-reading management, file management, personal information management, meeting management and other modules. It can realize non-paper office at the enterprise and institution, raise the office efficiency, spare the internal cost, and increase the economic benefits.

1. Production and Use Conditions

Consumers have the good hardware operation environment and the central server. Every department is equipped with the computers and the network environment of more than 100 = one hundred MHz.

2. Technical Traits

1) Universality: besides developing the special-purpose software for the special enterprises, the research and development of office automation software should aim at developing the general office software in order to be fit for the needs of all kinds of enterprises and institutions, as long as every consumer does the corresponding dispositions according to his own unit's specific reality, can achieve the office needs of his own unit, and realize the transition from the general purpose to the special purpose.

2) Automation: we can automate the repeated work in the office, the agency can be established, and administrators can own the unlimited Lotus Script agent competence on the server so that they operate the corresponding agency to realize the automation work. [1]

Regular statistics: we can count whether the units' internal leaders, organizations at different levels and members serve in time, and get the information about the document receiving, document sending, signature and statistics reporting, and other work, undertaking in time, overtime arrangement, and unfinished work overstock, etc..

Regular sending meeting plan: at the end of the week, a meeting plan for next week will be sent. This plan is made automatically for the attendants. When we make meeting arrangements, we should inquire the time if they are busy, and we can arrange for another time or rearrange persons selected.

Automatic deleting overdue document and files: when consumers work for a long time, there will be a lot of overdue documents and files overstock. Manual deletion can increase work load, and regular deletion agency is installed to delete overdue documents and files.

Automatic press: when the documents or some work in the units overstock and surpass the fixed time, the system automatically sends pressing notices to urge the staff concerned to dispose in time.

Business trip agency: when some staff is away on official business, the official business information should be set to "away". So when somebody sends this staff e-mail or document, the system will automatically notify the sender that this man is out, and whether still sending it or changing the person selected.

Automatically recording document circulating progress: all the process and procedures of document sending and document receiving can be automatically recorded and displayed to make the document flow clear.

Realizing groups sending: we can install the unit's internal different groups. When we send e-mail, send meeting notices, or send and receive documents, we can choose the corresponding groups. The system can automatically send them to all the members instead of sending them one by one. When there are many addressees, this way will greatly reduce the senders' work load.

The system automatically points out new received mail: when the staff is doing other work and the new mail is arriving or the document is to be dealt with, the system can automatically realize the visual and prompt notice that the staff should cope with them.

This mode is significant for the important documents to automatically affirm whether the mail is sent successfully.

3) Safety: in order to ensure the protection of the important information during the information exchange, different safety controlling can be setup levels so as to meet the consumers' safety demand.

As China entered the WTO, the Chinese enterprises will more face to the world competition. This not only demands the enterprises to strengthen the management, increase the production technique but also demands the enterprises to improve the internal efficiency and reduce the internal transition cost. Using network office workbench can realize no paper office, quickly transmit more information, and realize the teamwork.

New Words and Phrases

correspondence	[ˌkɒrɪˈspɒnd(ə)ns]	n. 信件，函件，通信，一致，相似
module	[ˈmɒdjuːl]	n. 单元，单位，（宇宙飞船上各个独立的）舱
enterprise	[ˈentəpraɪz]	n. 事业心，进取心，企〖事〗业单位
institution	[ˌɪnstɪˈtjuːʃ(ə)n]	n. 惯例，习俗，制度，慈善机构
trait	[treɪt]	n. 人的个性，显著的特点，特征
disposition	[ˌdɪspəˈzɪʃ(ə)n]	n. 气质，天性，性格，安排，布置
transition	[trænˈzɪʃ(ə)n]	n. 过渡，转变，变迁
competence	[ˈkɒmpɪt(ə)ns]	n. 能力，技能，（法院的）权限，管辖权
dispose	[dɪˈspəʊz]	vt. & vi. 处理，处置，布置
overdue	[ˌəʊvəˈdjuː]	adj. 迟到的，延误的，过期的，到期未付的
circulate	[ˈsɜːkjuleɪt]	vt. （使）循环，（使）流通，（使）流传，散布，传播
affirm	[əˈfɜːm]	vt. & vi. 断言，证实
conform	[kənˈfɔːm]	vi. 遵守，符合，顺应，一致
conform to	符合，遵照	

Technical Terms

file management	档案管理
non-paper office	无纸化办公

overdue document　过期文档
document flow　文档流程
transition cost　交易成本
network office workbench　网络工作平台

Notes

[1] We can automate the repeated work in the office, the agency can be established, …on the server so that they operate the corresponding agency to realize the automation work.

译文：我们可以将办公中需重复处理的工作实现自动化，可以建立代理，由管理员在服务器上拥有无限制的 Lotus Script 代理权限，运行相应的代理就可以实现自动化的工作。

说明：句中过去分词 repeated 作定语，can be established 是含有情态动词的被动语态形式，so that 引导的是一个结果状语从句。

Exercises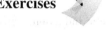

Ⅰ. Answer the following questions.

1. What's the function of Office Automation System?
2. What are the conditions of using Office Automation System?
3. How can we realize the automation work?
4. How can we ensure the safety of important information during the information exchange?

Ⅱ. Decide whether the statements are true (T) or false (F) according to the text.

1. Office Automation System helps to raise the office efficiency, spare the internal cost, and increase the economic benefits.
2. Although Office Automation System is used, we still need a lot of paperwork in the office.
3. The general office software is fit for any kind of enterprises and institutions.
4. With the help of Office Automation System, we can automate the repeated work in the office.

Ⅲ. Translate the following into English

1. 网络管理　　　2. 无纸化办公　　　3. 中央服务器
4. 会议安排　　　5. 个人信息　　　　6. 技术特性
7. 除了是位机械技师外，他还是位大诗人。
8. 第二种则认为，作为一个发展中国家，中国做的已经远远超出其公平份额。

Ⅳ. Translate the following paragraph.

Office automation system should achieve the following functions. The daily management of the enterprise should include conference management and announcement management business. Users can view business meetings and announcements published by the module, and set the function to add new announcements and meetings.

Translating Skills

定语从句的翻译

定语从句一直是科技英语阅读中经常遇到却又难以把握的内容。定语从句一般位于所修饰的名词（先行词）后，或与先行词有一定的间隔。定语从句究竟修饰文中的哪个词，一要靠语言知识理解，二要根据上下文进行逻辑判断。

定语从句可以分为限制性定语从句和非限制性定语从句两种。限制性定语从句对修饰的先行词起限制作用，在意义上与先行词密不可分，是主句意义中不可缺少的成分，一般不可用逗号与主句分开；而非限制性定语从句仅对先行词在原因、让步方面进行解释，或对时间、条件、结果等进行补充说明，关系代词不能省略，从句与主句用逗号分开，引导词有which（不是that）、when、who 等。

定语从句的一般结构是先行词+引导词（包括关系代词 who、that、which、whom、whose 和关系副词 when、where、why、how）引出的从句。

A transformer is composed of some coils that are coupled together by magnetic induction. 变压器由磁感应耦合线圈组成。（限制性定语从句）

ISDN is a new service offered by many telephone companies that provides fast, high-capacity digital transmission of voice, data, still images and fullmotion video over the worldwide telephone network.

综合业务数据网是由电话公司提供的新型业务，该业务可在世界范围内的电话网上进行语音、数据、静止图像和全动态视频的快速、大容量数字传输。（限制性定语从句）

When the pointer stops moving, the reading given by the ohmmeter is the insulation resistance, which is normally high if the capacitor is in good condition.

指针停止移动时的读数就是电容器的绝缘电阻值，假如电容器正常工作，绝缘电阻值通常很高。（非限制性定语从句）

First of all, let's introduce you to some of the circuits that are commonly found in radio-frequency systems and with which you may not be familiar.

首先让我们向你介绍射频系统中常见的一些电路，或许你并不熟悉它们。（介词+which 构成从句，类似的句型还有介词+whom 等）

The analog world is full of relative numbers, tradeoffs, and approximations, all of which depend heavily on the basic semiconductor properties.

模拟电子技术领域充满了相对数、折中和近似值，所有这些在很大程度上取决于半导体的基本特性。（代词+介词+which 构成从句，句中的 all 指代前面的数值等）

Direct current is an electric current the charges of which move in one direction only.

直流电是电荷只向一个方向流动的电流。（名词+介词+which 构成从句）

注意：

1) 在限制性定语从句中，引导词指代先行词，定语从句的引导词在从句中充当某个成分，有时可将其省略。例如：

The GPS receiver compares the time a signal was transmitted by a satellite with the time it was received. GPS 接收机把卫星发送信号的时间和接收信号的时间相比较。

80

2）在非限制性定语从句中不能使用 that 作为关系代词，而只能用 who、whom、which。例如：

Microwaves, which have a higher frequency than ordinary radio waves, are used routinely in sending thousands of telephone calls and television programs across long distance. 微波比一般无线电波信号的频率更高，它通常被用来进行远距离传送大量的长途电话和电视节目。（定语从句解释原因）

Reading

Multimedia Technology

You know you have good reason to own a business strength PC at home. It helps you to complete the work chores that would otherwise keep you staying up late at the office. [1] But what can you do with your home PC when you are not working? That is where multimedia comes in. It is relatively cheap and easy to add stereo sound, a CDROM drive and Windows to today's work at home PC. What you end up with, though, is something far more revolutionary than just a computer that talks, plays compact discs, and displays pretty pictures. [2] Your multimedia PC (MPC, for short) is the key to discover books and encyclopedias that bring ideas to life through images and speechs, games that match the quality and fun of the best entertainment, and software that unleashes your personal creativity in music, animation, and video.

Multimedia is a computer technology. Its applications involve the integrated processing of various types of data, like sounds, images, full motion videos, graphics, character data, etc. Now, multimedia applications have been developed for traditional uses like customer service, office automation (OA), and computer aided instruction (CAI). In the multimedia environment, we have graphics and texts at the same time, we can also add photographs, animations, good-quality sounds, and full motion videos. Development of the technologies make computers much more powerful and much easier to use.

The requirements for a MPC are described in detail in Microsoft's Multimedia PC Specification Version 10. Here are the key terms of the specifications.

Full motion video: full motion video is digitally recorded video played back at the broadcast standard of 30 frames per second, or closes enough to that speed, so the video appears smooth rather than jerky.

MIDI: MIDI is short for Musical Intranet Digital Interface, a standard specification developed by music synthesizer manufacturers. [3] The concept of being able to control several instruments from one keyboard has grown into a method for putting musical instruments, tape recorders, VCRs, mixers, and even stage lighting under the control of a single computer.

New Words and Phrases

multimedia [ˈmʌltɪmiːdiə] adj. 多媒体的 n. [计] 多媒体
chore [tʃɔː] n. 家庭杂务；日常的零星事务；讨厌或累人的工作

stereo	[ˈsteriəu]	adj. 立体的，立体声的 n. 立体声；立体声系统
compact disc		光盘
encyclopedia	[inˌsaikləˈpi:diə]	n. 百科全书
unleash	[ʌnˈli:ʃ]	vt. 发动；解除……的束缚 vi. 不受约束；自由自在
involve	[inˈvɒlv]	vt. 包括；涉及；使陷于；潜心于
animation	[ˌæniˈmeiʃən]	n. 活泼；生气；激励；卡通片绘制
specification	[ˌspesifiˈkeiʃən]	n. 规格；说明书；详述
jerky	[ˈdʒɜːki]	adj. 急动的；不平稳的
synthesizer	[ˈsinθisaizə]	n. [电子] 合成器；合成者

Notes

[1] It helps you to complete the work chores that would otherwise keep you staying up late at the office.

译文：它能帮助你完成日常工作琐事，否则你会被困在办公室里被迫熬夜。

说明：that 引导限制性定语从句，修饰前面的名词 work chores。keep sb doing sth 是使某人一直做某事，说明动作的持续性。Staying up late 翻译为"熬夜，开夜车"。

[2] What you end up with, though, is something far more revolutionary than just a computer that talks, plays compact discs, and displays pretty pictures.

译文：虽然你最终得到的收获远比一台仅能说话、能播放光盘和能显示美丽图像的计算机更具有革命性。

说明：what 引导主语从句，谓语动词是 is。end up with 意思是"以……告终"。far more than 意思是"远远多于，远不止于"，其用法和 more than 相同，只是程度上有所增加。

[3] MIDI is short for Musical Intranet Digital Interface, a standard specification developed by music synthesizer manufacturers.

译文：MIDI 是乐器数字接口的缩写，它是由音乐合成器制造商开发的标准规范。

说明：is short for 意为"是……的简称，缩写"。developed by 是过去分语短语作定语，修饰前面的 a standard specification，而 a standard specification 是 Musical Intranet Digital Interface 的同位语。

Exercises

I. Answer the following questions.
1. What's the good reason for you to own a business strength PC at home?
2. What are the revolutionary changes a multimedia PC can bring to you?
3. What is the application of multimedia?
4. Why does the full motion video appear smooth rather than jerky?

II. Match the following phrases in column A with Column B.

Column A   Column B
1. multimedia personal computer   a. 光驱，光碟机
2. CDROM drive   b. 全自动录像

3. office automation   c. 计算机辅助教学
4. full motion video   d. 多媒体个人计算机
5. computer aided instruction   e. 办公自动化
6. customer service   f. 客户服务

Unit 13 Security System

Text

With the rapid development of science and technology, people have higher and higher security requirements in habitation, living and working environment. It is an urgent need for a safe, reliable, security protection system to protect people's lives and property safety. So the intelligent security system is invented to satisfy the people's security concerns.

Security technology system, such as digital closed-circuit television monitoring system and burglar alarm system is an important component of the intelligent system. It is an essential subsystem of the intelligent system. It is an integrated system with advanced and excellent ability of safety precautions. Through the remote-controlled cameras and its auxiliary, it can directly watch the place which is inconvenient for people to observe by eyes. [1] It provides a real-time, vivid and economic monitoring way. At the same time it can record some parts or all the parts of the monitoring areas. For that reason, it provides the convenience condition or important basis for handling some things in the future. The digital closed-circuit television monitoring system can also be linked with anti-theft and other security technology system to make the ability of safety precautions more powerful. Therefore, it has been widely used in the industrial, residential, hotel and other areas.

The system of the security technology equipment, facilities and its components can make rapid response to intruders and detect or capture criminals in time. It has a strong deterrent to the criminals. The safety precautions technology can promptly find the hidden dangers of accidents, prevent damage, and reduce accidents and prevent disaster, therefore it is a very important means of safe working. Especially in the today's highly developed modern technology, crime is more intelligent, more subtle, so strengthening the modernization of security technology has become more important.

In general, the security system includes closed-circuit monitoring system, building intercom system, burglar alarm system, intelligent home system, import and export control system, fire alarm system etc..

Closed-circuit monitoring system is the most important sub-system, which carries out security surveillance and management of surveillance through laying out a certain number of cameras at the scene. [2]

Building intercom system is a modern property designed for the security system. [3] It combines indoor security alarm system with the entrance and exit control system as a whole. It can realize the call, intercom, video surveillance, security alarm, unlock, three call functions etc. to achieve the purpose of intelligent property management.

Anti-theft alarm system lays out the mobile detectors, infrared detectors, door magnetic detector, glass breaking detectors, smoke detectors, emergency buttons and other detectors in the scene, when the illegal invasion happens or in case of any changes it will issue a warning signal so as to effectively protect people's lives and property.

Entrance and exit control system usually refers to the use of modern electronics and information technology in the entrance and exit to aim at people, control their release, refuse, record and alarm the operation systems, including access control, car park management system.

Intelligent home security system is an important component of the security system, and through the corresponding communication host, home alarm will be combined with the remote control, not only with security, control, alarm function, but also, by telephone or mobile phone, to realize remote control home's any kind of electrical appliances' on-off.

Security and intelligence are the inevitable trend of the social informatization. Nowadays high-tech crime cases are increasing, security system shows the backbone role of security guards, such as closed-circuit television monitoring system, perimeter guard system, anti-theft alarm system, building intercom system, smart parking management system, public address system and os on. Statistics show that the installation of a building security system reduces the crime rate by about 80%.

New Words and Phrases

security　　　[sɪˈkjuərəti]　*n.* 安全；保证；证券；抵押品　*adj.* 安全的；保密的
habitation　　[ˌhæbɪˈteɪʃ(ə)n]　*n.* 居住；居所
burglar　　　[ˈbɜːglə]　*n.* 窃贼；破门盗窃者
precaution　　[prɪˈkɔːʃ(ə)n]　*n.* 预防，警惕；预防措施　*vt.* 警惕；预先警告
auxiliary　　　[ɔːgˈzɪliəri]　*n.* 辅助者；辅助物　*adj.* 辅助的；副的；附加的
residential　　[ˌrezɪˈdenʃ(ə)l]　*adj.* 住宅的；与居住有关的
facility　　　[fəˈsɪləti]　*n.* 设施；设备
intruder　　　[ɪnˈtruːdə]　*n.* 侵入者；干扰者；闯入者（尤指企图行窃者）
capture　　　[ˈkæptʃə]　*vt.* 俘获；捕捉；拍摄；录制　*n.* 战利品；俘虏
detect　　　[dɪˈtekt]　*vt.* 察觉；发现；探测
deterrent　　　[dɪˈterənt]　*adj.* 威慑的；遏制的　*n.* 威慑；妨碍物
subtle　　　[ˈsʌt(ə)l]　*adj.* 微妙的；精细的；敏感的；狡猾的；稀薄的
intercom　　　[ˈɪntəkɒm]　*n.* 对讲机；内部通话装置
surveillance　　[səˈveɪl(ə)ns]　*n.* 监督，监视
property　　　[ˈprɒpəti]　*n.* 性质；性能；财产；所有权
infrared　　　[ˌɪnfrəˈred]　*n.* 红外线　*adj.* 红外线的
magnetic　　　[mægˈnetɪk]　*adj.* 地磁的；有磁性的；有吸引力的
invasion　　　[ɪnˈveɪʒ(ə)n]　*n.* 入侵；侵略；侵袭；侵犯
issue　　　　[ˈɪʃuː]　*n.* 问题；流出；发行物　*vt. & vi.* 发行；发布
appliance　　　[əˈplaɪəns]　*n.* 器具；器械；装置；电器
inevitable　　　[ɪnˈevɪtəb(ə)l]　*adj.* 必然的；不可避免的
informatization　[ɪnfɔːmətaɪˈzeɪʃn]　*n.* 信息化
link with　　与……有关，与……相连接
carry out　　执行；实施；贯彻；实现；完成
lay out　　　设计；展示；安排；陈设
in case of　　万一；如果发生；假设

Technical Terms

closed-circuit television monitoring system 闭路电视监控系统
remote-controlled camera 远程控制摄像机
burglar alarm system 防盗报警系统
building intercom system 楼宇对讲系统
intelligent home system 智能家居系统
import and export control system 进出口门禁控制系统
fire alarm system 消防报警系统
smart parking management system 智能停车场管理系统
public address system 公共广播系统

Notes

[1] Through the remote-controlled cameras and its auxiliary, it can directly watch the place which is inconvenient for people to observe by eyes.

译文：通过遥控摄像机及其辅助设备，它能直接看到人类肉眼不便观测到的地方。

说明：句中 where 引导的限制性定语从句修饰 the place。限制性定语从句对被修饰的词有限定制约作用，使该词的含义更具体，在句中不能省略，否则句意不完整。

[2] Closed-circuit monitoring system is the most important sub-system, which carries out security surveillance and management of surveillance through laying out a certain number of cameras at the scene.

译文：闭路监控系统是安防系统最重要的子系统，它通过在现场布置一定数量的摄像机进行安全监控和监控管理。

说明：句中 which 引导的是非限制性定语从句，which 指代前面的 closed-circuit monitoring system。非限制性定语从句与先行词的关系并不是非常密切，只是作为附加说明，去掉后主句意义仍然完整。

[3] Building intercom system is a modern property designed for the security system.

译文：楼宇对讲系统是为现代物业设计的安全系统。

说明："designed for…" 在句中是过去分词短语作定语，修饰前面的名词 "a modern property"，相当于定语从句 "which is designed for…"。意思是 "为……设计的"。

Exercises

I. Answer the following questions.

1. What provides the important basis for handling some things in the future?
2. What does the security system include?
3. What is the function of building intercom system?
4. How does the intelligent home security system realize remote control home's any kind of electrical appliances' on-off?

Ⅱ. Decide whether the statements are true (T) or false (F) according to the text.

1. The digital close-circuit television monitoring system can record some parts or all the parts of the monitoring areas.

2. Security system can not prevent damage and reduce accidents.

3. Building intercom system can realize the call, intercom, video surveillance, security alarm, unlock, three call functions, etc.

4. Intelligent securit system is the inevitable trend of the social informatization.

Ⅲ. Translate the following phrases into English.

1. 安全技术体系　　2. 数字闭路电视监控系统　　3. 防盗报警系统
4. 通信主机　　　　5. 楼宇对讲系统　　　　　　6. 智能家居系统
7. 门禁控制系统　　8. 远程控制

Ⅳ. Translate the following sentences into Chinese.

1. With the rapid development of science and technology, people have higher and higher security requirements in habitation, living and working environment.

2. Close-circuit monitoring system carries out security surveillance and management of surveillance through laying out a certain number of cameras at the scene.

3. Security and intelligence are the inevitable trend of the social informatization.

4. Statistics show that the installation of a building security system reduces the crime rate by about 80%.

Translating Skills

虚拟语气的翻译

科技文章中常常提出一些设想、推理或判断,内容与事实相反,或者不大可能实现。为了同客观存在的实际相区别,要求用虚拟语气。

1. 简单虚拟句

简单虚拟句由 should/would/could/might + 动词原形构成,表示主观对客观事物的看法、愿望、请求、建议等。

The receiver <u>should recover</u> the baseband signal exactly. 接收机应完全恢复基带信号。

In this case, <u>there would be</u> no need to have security mechanisms within the network itself. 在这种情况下,网络本身不需要建立密码机制。

2. 虚拟语气用于 if 条件句

If there <u>were</u> no attraction between the proton and the electron, the electron <u>would fly</u> away from the proton in a straight line. 倘若质子与电子间不存在引力,电子就会沿直线飞离质子。(表示与现在事实相反)

The bit stream that we finally arrive at is smaller than what <u>would</u> have resulted for using ASCII/UNICODE Tables. 最终我们所得到的比特流,比我们采用 ASCII/UNICODE 得到的要小。(表示与过去事实相反)

如果上述条件状语从句中省略 if,句子的主语和谓语动词部分要倒装。

If we <u>were to travel</u> by space rocket to the moon, we <u>would found</u> it quite a different place from our own planet—the earth. 若我们乘宇宙飞船到月球上去旅行，将会发现，月球与我们自己的星球——地球是完全不一样的。（表示与将来事实相反）

3. 虚拟语气用于宾语从句

用在 desire、demand、suggest、recommend、advise、decide、order 等表示"愿望、建议、命令、要求、忠告"的动词后面的宾语从句中往往用"should + 动词原形"或动词原形的虚拟语气。

High efficiency implies that circuit loss be minimum and the ratio of the transistor output, the parallel equivalent resistance, and its collector load resistance be maximum. 高效率意味着电路损耗应最小，而晶体管输出比率、并联等效电阻及集电极负载电阻应最大。

Communication over long distances usually requires that some alterations or other operations be performed on the electrical signal conveying the information in preparation for transmission. 远距离通信，通常需要对拟传输的承载信息的电信号进行某些变换或操作。

4. 虚拟语气用于主语从句

"It is（was）+形容词（或过去分词、名词）+ that…"结构中的虚拟语气，其表达形式为 should + 动词原形或省略 should 直接用动词原形。

常用形容词有 appropriate、advisable、preferable、necessary、important、imperative、urgent、essential、vital、possible、compulsory、crucial 等。

常用过去分词有：required、demanded、desired、suggested、ordered 等。

常用名词有：advice、decision、desire、demand、order、preference、proposal、recommendation、requirement、resolution、suggestion 等。

It is necessary that the theoretical sections of the book be carefully studied and thoroughly understood. 必须仔细研究并透彻理解本书的理论部分。

It is my proposal that he study the information theory first. 我建议他应该先学信息论。

5. 虚拟语气用于状语从句

The sum of the kinetic energy and potential energy of the system is always the same, provided the system be not acted upon by anything outside it. 假定系统不受任何外力作用，系统的动能和势能之和就始终保持不变。（表示条件）

Audio signals are attenuated very rapidly, so we would have to be within a few hundred feet of the originating source in order to hear the signal. 音频信号衰减很快，所以要在距离声源几百英尺范围以内才能听到它。（表示结果）

6. 虚拟语气用于定语从句

A computation which would have taken years of human work in the past, is now done in a few seconds with the help of computers. 过去需要人工作几年的运算，现在借助计算机几秒钟就完成了。

Reading

Intelligent City

Intelligent city centers on urbanization development, sustainable city growth and core demands of urban residents. Intelligent city requires effective integration of advanced information technology with advanced operating and service philosophy[1] (as shown in Fig. 13-1). An intelligent city will collect and store a multitude of a city's information resources in real time to create its IT infrastructure, and by data interconnection and interoperability, exchange and sharing, and collaborated applications, it will create a platform which provides a convenient, efficient, and flexible tool for generating and implementing decisions related to the city management and operation, as well as for the provisioning and management of innovative public services[2], with an ultimate goal of achieving harmonious development of safer, greener, more efficient and more convenient urbanization.

Fig. 13-1　Intelligent City

Every smart city should have is presented with examples of applications of these in different cities. The list of the technologies is the following:

1. Open-data initiatives and hackathons
2. Parking apps
3. Apps that let users "adopt" city property
4. High-tech waste management systems
5. All-digital and easy-to-use parking payment systems
6. A city guide app
7. Touchscreens around the city
8. Wi-Fi in subway stations and on trains
9. Sustainable and energy efficient residential and commercial real estate
10. Dynamic kiosks that display real-time information
11. App or social media-based emergency alert and crisis response systems
12. Police forces that use real-time data to monitor and prevent crime
13. More public transit, high-speed trains, and bus rapid transit (BRT)
14. OLED lights and surveillance in high-crime zones
15. Charging stations, like the solar-powered
16. Roofs covered with solar panels or gardens
17. Bike-sharing programs
18. A sharing economy, instead of a buying economy
19. Smart climate control systems in homes and businesses
20. Widespread use of traffic rerouting apps
21. Water-recycling systems
22. Crowdsourced urban planning
23. Broadband Internet access for all citizens
24. Mobile payments
25. Ride-sharing programs

New Words and Phrases

sustainable [səˈsteɪnəbl] *adj.* 可持续的；合理利用的
integration [ˌɪntɪˈɡreɪʃn] *n.* 集成；综合；同化
multitude [ˈmʌltɪtjuːd] *n.* 大量；多数；群众
infrastructure [ˈɪnfrəstrʌktʃə(r)] *n.* 基础；基础设施
interconnection [ɪntəkəˈnekʃn] *n.* 互相联络
interoperability [ˌɪntərɒpərəˈbɪləti] *n.* 互操作性；互用性
hackathon [ˈhɑːkəθɒn] *n.* 黑客马拉松；编程马拉松
urbanization [ˌɜːbənaɪˈzeɪʃn] *n.* 都市化；文雅化
crowdsource [ˈkraʊdˌsɔːs] *vi.* 众包（尤指用互联网将工作分给不特定人群），群众外包
surveillance [səˈveɪl(ə)ns] *n.* 监督；监视
kiosks [ˈkiːɒsks] *n.* 报摊，电话亭（kiosk的复数）；店铺设计
solar-powered [ˈsəʊləˈpaʊəd] *adj.* 太阳能的
collaborate [kəˈlæbəreɪt] *v.* 合作；勾结；协调
implement [ˈɪmplɪmənt] *n.* 工具；器具 *vt.* 履行；实施
harmonious [hɑːˈməʊniəs] *adj.* 和蔼的；和睦的；音调优美的。
dynamic [daɪˈnæmɪk] *adj.* 动力的；动态的；有活力的 *n.* 动力；动力学
sustainable city growth　城市可持续发展
water-recycling system　水循环系统
smart climate control system　智能温控系统
sharing economy　分享型经济
buying economy　购买型经济
solar panel　太阳能电池板
OLED　有机发光二极管
people communities　居民社区
natural environment　生态环境保护
built infrastructure　基础设施服务
smart city initiative　智能城市规划

Notes

[1] Intelligent city requires effective integration of advanced information technology with advanced operating and service philosophy.

译文：智能化城市需要有效地融合先进的信息技术和先进的操作服务理念。

说明：integrate with 意思是"使与……结合" 又如：

integrate methodology with practice 理论和实践相结合。

句中 integration 是 integrate 的名词形式，意思是"集成；综合"。名词化是科技英语的特点之一，也是正式文体的最显著的特征之一。在翻译的时候我们可以把名词还原成动词。

[2] It will create a platform which provides a convenient, efficient, and flexible tool for generating and implementing decisions related to the city management and operation, as well as for the provisioning and management of innovative public services.

译文：它将创建一个平台，这个平台提供一种方便、高效、灵活的工具用于生成和实现与城市管理和操作相关的决策，以及配置和管理创新公共服务。

说明：which 引导定语从句，修饰前面的名词 platform。

as well as 是并列连词，用来连接两个并列的成分，作"也，还"解。它强调的是前一项，后一项只是顺便提及。因此连接并列主语时，谓语动词与前一项一致。如：

Electric energy can be changed into light energy as well as into sound energy. 电能既可以被转变成声能，又可以被转变成光能。

Exercises

I. Answer the following questions.

1. What does intelligent city focus on?

2. In which way does intelligent city create a platform which provides a convenient, efficient, and flexible tool?

3. What is the ultimate goal of intelligent city?

4. What is an intelligent city like in your mind?

II. Translate the following sentences into Chinese.

1. Intelligent city requires effective integration of advanced information technology with advanced operating and service philosophy.

2. An intelligent city will collect and store a multitude of a city's information resources in real time to create its IT infrastructure.

3. It will create a platform which provides a convenient, efficient, and flexible tool for generating and implementing decisions related to the city management and operation.

Chapter III   Automatic Control Technology

Unit 14   Introduction to Control Engineering

Text

In recent years there has been a considerable advance made in the art of automatic control. The art is, however, quite old, stemming back to about 1790 when James Watt invented the centrifugal governor to control the speed of his steam engines. He found that while in many applications an engine speed independent of load torque was necessary, in practice when a load was applied the speed fell and when the load was removed the speed increased.

In a simple centrifugal governor system, variations in engine speed are detected and used to control the pressure of the steam entering the engine. Under steady conditions the moment of the weight of the metal spheres balances that due to the centrifugal force and the steam valve opening is just sufficient to maintain the engine speed at the required level.[1] When an extra load torque is applied to the engine, its speed will tend to fall, the centrifugal force will decrease and the metal spheres will tend to fall slightly. Their height controls the opening of the steam valve which now opens further to allow a greater steam pressure on the engine. The speed thus tends to rise, counteracting the original tendency for the speed to fall. If the extra load is removed, the reverse process takes place, the metal spheres tend to rise slightly, so tending to close the steam valve and counteracting any tendency for the speed to rise.

It is obviously that without the governor the speed would fall considerably on land. However, in a correctly designed system with a governor the fall in speed would be very much less. An undesirable feature which accompanies a system which has been designed to be very sensitive to speed changes is the tendency to "hunt" or oscillate about the final speed. The real problem in the synthesis of all systems of this type is to prevent excessive oscillation but at the same time produce good "regulation". Regulation is defined as the percentage change in controlled quantity on load relative to the value of the controlled under condition of zero load. Regulators form an important class of control system, their object generally being to keep some physical quantity constant (e. g. speed, voltage, liquid level, humidity, etc.) regardless of load variation. A good regulator has only very small regulation.

The 1914~1918 war caused military engineers to realize that to win wars it is necessary to position heavy masses (e. g. ships and guns) precisely and quickly.[2] Classic work was performed by N. Minor sky in the USA in the early 1920s on the automatic steering of ships and the automatic positioning of guns on board ships. In 1934 the word "servomechanism" (derived from the Latin) was used in the literature for the first time by H. L. Hazen. He defined a servomechanism as "a power amplifying device in which the amplifier element driving the output is actuated by the difference between the input to the servo and its output". This definition can be applied to a wide variety of "feedback control systems". More recently it has been suggested that the term "servomechanism" or "servo" be restricted to a feedback control system in which the controlled variable is mechanical position.[3]

The automatic control of various large-scale industrial processes, as encountered in the manufacture and treatment of chemicals, food and metals, has emerged during the last thirty years as an extremely important part of the general field of control engineering. In the initial stages of development it was scarcely realized that the theory of process control was intimately related to the theory of servomechanisms and regulators. Even nowadays complete academic design of process control systems is virtually impossible owing to our poor understanding of the dynamics of processes. In much of the theory introduced in this book, servomechanisms and regulators are used as examples to illustrate the methods of analysis.

New Words and Phrases

    sphere   [sfɪə]   *n.* 范围，球体   *vt.* 放入球内，使……成球形   *adj.* 球体的
    counteract   [ˌkaʊntəˈrækt]   *vt.* 抵消，中和，阻碍
    centrifugal   [senˈtrɪfjugəl]   *adj.* 离心的，远中的   *n.* 转筒，离心机
    accompany   [əˈkʌmpənɪ]   *vt.* 陪伴，伴随，伴奏   *vi.* 伴唱
    encounter   [ɪnˈkaʊntə]   *vt.* 遭遇，邂逅，遇到   *vi.* 偶然碰见
    dynamics   [daɪˈnæmɪks]   *n.* 动力学，力学
    steer   [stɪə]   *vt.* 驾驶，控制，引导   *vi.* 掌舵，行驶
    servomechanism   [ˌsɜːvəʊˈmekənɪzəm]   *n.* 伺服机构，自动控制装置
    independent of   不受……支配的，与……无关的
    be sensitive to   对……敏感

Technical Terms

    centrifugal governor   离心式调速器
    feedback control system   反馈控制系统
    the theory of process control   过程控制理论

Notes

[1] Under steady conditions the moment of the weight of the metal spheres balances that due to the centrifugal force and the steam valve opening is just sufficient to maintain the engine speed at the required level.

译文：在稳定条件下，由于离心力的作用和蒸汽阀的开度刚好足够维持发动机转速所要求的水平，瞬间与金属摆球重量平衡。

说明：句中 is just sufficient to maintain（刚好足以维持），at the required level（以所需要

的量），其中 balance 译作"保持平衡；相称；抵消"，又如：

How do you optimally balance a thousand counteracting variables in a complex problem? 如何最有效地平衡复杂问题中的 1000 种对抗的变数？

［2］The 1914 ~ 1918 war caused military engineers to realize that to win wars it is necessary to position heavy masses（e. g. ships and guns）precisely and quickly.

译文：1914 ~ 1918 年战争促使军队工程师意识到为了赢得战争胜利需要准确而迅速地使重型装备（例如船只和枪炮）定位。

说明：动词不定式 to realize 表示目的，that 引导宾语从句。从句中 it 是形式主语，to position heavy masses… 是真正主语。position 作动词，译为"定位"，又如：

The depositor positively positioned the preposition in that position on purpose. 存款人明确地有目的地把介词放到那个位置。

［3］More recently it has been suggested that the term "servomechanism" or "servo" be restricted to a feedback control system in which the controlled variable is mechanical position.

译文：最近更多人提议伺服机构和伺服受到反馈控制系统的机械位置变量的限制。

说明：it has been suggested that…（有人建议……），that 引导主语从句，suggest 后面的从句用虚拟语气，be 动词用原型。be restricted to，限制于。

又如：Time has now come to the necessary legislation to be restricted to this moment. 现在已到了需要立法来对此加以限制的时机。

Exercises

I . Answer the following questions according to the text.

1. What is the function of the centrifugal governor?
2. What is the problem in the sensitive speed changes system?
3. What is the function of the regulator?
4. In the early 19th century, what is the definition of the servomechanism?

II. Fill in the blank.

1. The steam valve opening is just sufficient to maintain the engine speed _____ the required level.
2. When James Watt _____ the centrifugal governor to control the speed of his steam engines.
3. When a load was applied the speed fell and when the load was removed the speed _____.
4. In a simple centrifugal governor system, variations in engine speed are _____ and used to control the pressure of the steam entering the engine.

III. Translate the following sentences into English.

1. 显然，去掉负荷发动机的转速会上升很快。
2. 由于离心力的存在，金属球刚好足以维持发动机转速所需要的水平。
3. 在研究的初期很少有人认识到控制理论的重要性。
4. 调节作用被定义为负载条件下被控制量的数值相对空载条件下被控制量数值的变化百分比。

Translating Skills

and 引导的句型的译法

and 作为连词,用来连接词、短语和句子,其基本意义相当于汉语的"和""与""并且"。但在实际翻译的过程中,特别是在连接两个句子时,它的译法很多,表达的意义可能相差甚远。如果不考虑 and 前后成分之间的逻辑关系,只用几种基本译法生硬套用,就难免造成理解上的失误,甚至把整个句子意思搞错。

1)表示原因,例如:
Laser is widely used for developing many new kinds of weapons, and it penetrates almost everything. 激光广泛用于制造各种新式武器,因为为它的穿透力很强。

2)表示因果,例如:
But since a digital signal is made up of a string of simple pulses, noise stands out and easily removed. 但由于数字信号由一组简单脉冲组成,杂音明显,因而容易排除。

3)表示目的,例如:
It was later shown that the results of this work were by no means the ultimate, and further work has been put in hand and to provide closer control and more consistent operation in this area. 后来发现,这项研究工作的结果绝非已作定论,而且进一步的研究工作已开始,以便在这方面提供较严密的控制和持续的操作。

4)表示承接,例如:
In many ways, computer is more superior than human brain, and human can rule it.
在许多方面计算机超过人脑,而人却可以控制它。

5)表示对照,例如:
Motion is absolute, and stagnation is relative. 运动是绝对的,而静止是相对的。

6)表示递进,例如:
The electronic brain calculates a thousand times quicker, and more accurately than is possible for the human being. 计算机的运算速度比人所能达到的速度要快 1000 倍,甚至更加准确。

7)表示转折,例如:
There will always be some things that are wrong, and that is nothing to be afraid of.
错误在所难免,但并不可怕。

8)表示条件,例如:
Even if a programmer had endless patience, knowledge and foresight, storing every relevant detail in a computer, the machine's information would be useless, and the programmer knew little how to instruct it in what human beings refer to as commonsense reasoning. 即使一个编程员很有耐心、知识和远见,把每一个有关细节都存入计算机,如果他不懂得按人类常识推理去对计算机下达指令,那么机器里的信息也还是没有用途的。

9)表示结果,例如:
Operators found that the water level was too low so they turned on two additional main coolant pumps, and too much cold water flowing into the system caused the steam to condense, further destabilizing the reactor. 操作人员发现冷却水的水位过低,就启动了另外两台主冷却泵,结果

过量的冷却水进入系统使蒸汽冷凝，反应堆因而更不稳定。

Reading

Closed-Loop Control System

It is well known that there are three basic components in a closed-loop control system. They are:

1. The Error Detector

This is a device which receives the low-power input signal and the output signal which may be of different physical natures, converts them into a common physical quantity for the purposes of subtraction, performs the subtraction, and gives out a low-power error signal of the correct physical nature to actuate the controller. The error detector will usually contain "transducers"; these are devices which convert signals of one physical form into another.

2. The Controller

This is an amplifier which receives the low-power error signal, together with power from an external source. A controlled amount of power (of the correct physical nature) is then supplied to the output element.

3. The Output Element

It provides the load with power of the correct physical nature in accordance with the signal received from the controller. Other devices such as gear-boxes and "compensating" devices are often featured in control systems, but these can usually be considered to form part of one of the other elements.

New Words and Phrases

  detector [dɪˈtektə] *n.* 检测器，侦察器，发现者
  subtraction [səbˈtrækʃən] *n.* 减少，减法，差集
  transducer [trænzˈdjuːsə] *n.* 传感器，变换器，换能器
  compensate [ˈkɒmpenseɪt] *vi.* 补偿，赔偿，抵消，*vt.* 付报酬

Exercises

Translate the following sentences into Chinese.

1. It is well known that there are three basic components in a close-loop control system.

2. The error detector will usually contain "transducers"; these are devices which convert signals of one physical form into another.

3. It provides the load with power of the correct physical nature in accordance with the signal received from the controller.

Unit 15 Programmable Logic Controller (PLC)

Text

A programmable logic controller, PLC, or programmable controller is a small computer used for automation of real-world processes, such as control of machinery on factory assembly lines. [1] The PLC usually uses a microprocessor. The program can often control complex sequencing and is often written by engineers. The program is stored in battery-backed memory and/or E^2PROMs.

PLCs have the basic structure, shown in Fig. 15-1.

From the figure, the PLC has four main units: the programme memory, the data memory, the output devices and the input devices. The Programme memory is used for storing the instructions for the logical control sequence. The status of switches, interlocks, initial values of data and other working data is stored in the data memory.

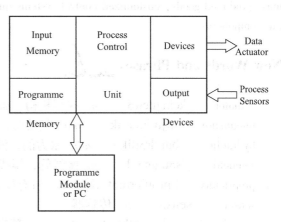

Fig. 15-1 Schematic of a PLC

The main differences from other computers are the special input/output arrangements. These connect the PLC to sensors and actuators. PLCs read limit switches, temperature indicators and the positions of complex positioning systems. Some even use machine vision. On the actuator side, PLCs drive any kind of electronic motor, pneumatic or hydraulic cylinders, diaphragms, magnetic relays or solenoids. The input/output arrangements may be built into a simple PLC, or the PLC may have external I/O modules attached to a proprietary computer network that plugs into the PLC. [2] PLCs were invented as less expensive replacements for older automated systems that would use hundreds or thousands of relays and cam timers. Often, a single PLC can be programmed to replace thousands of relays. Programmable controllers were initially adopted by the automotive manufacturing industry, where software revision replaced the rewiring of hard-wired control panels.

The functionality of the PLC has evolved over the years to include typical relay control, sophisticated motion control, process control, distributed control systems and complex networking. The earliest PLCs expressed all decision making logic in simple ladder logic inspired from the electrical connection diagrams. The electricians were quite able to trace out circuit problems with schematic diagrams using ladder logic. This was chosen mainly to reduce the apprehension of the existing technicians.

Today, the PLC has been proven very reliable, but the programmable computer still has a way to go. With the IEC 61131-3 standard, it is now possible to program these devices using structured programming languages, and logic elementary operations.

A graphical programming notation called Sequential Function Charts is available on certain programmable controllers.

However, it should be noted that PLCs no longer have a very high cost (often thousands of dollars), typical of a "generic" solution. Modern PLCs with full capabilities are available for a few hundred USD. There are other ways for automating machines, such as a custom microcontroller-based design, but there are differences between both: PLCs contain everything needed to handle high power loads right out of the box, while a microcontroller would need an electronics engineer to design power supplies, power modules, etc. Also a microcontroller based design would not have the flexibility of in-field programmability of a PLC. That is why PLCs are used in production lines, for example. These typically are highly customized systems so the cost of a PLC is low compared to the cost of contacting a designer for a specific, one-time only design. On the other-hand, in the case of mass-produced goods, customized control systems quickly pay for themselves due to the lower cost of the components.

New Words and Phrases

  actuator   [ˈæktjueitə]   n. 执行机构,激励者,促动器
  pneumatic   [njuːˈmætik]   adj. 气动的,有气胎的,充气的   n. 气胎
  hydraulic   [haiˈdrɔːlik]   adj. 水力的,液压的,水力学的
  solenoid   [ˈsəulənɔid]   n. 螺线管,螺线形电导管
  proprietary   [prəuˈpraiətəri]   n. 所有权,所有人   adj. 所有的,专利的
  sensor   [ˈsensə]   n. 传感器
  apprehension   [æpriˈhenʃ(ə)n]   n. 理解,逮捕,恐惧,忧惧
  module   [ˈmɔdjuːl]   n. 模数,模块,组件
  trace out   描绘出
  oscillate   [ˈɔsileit]   vt. 使动摇,使振荡,vi. 振荡,摆动

Technical Terms

  battery-backed memory   随机存储器
  E²PROM   电可擦写可编程序只读存储器
  Distributed Control Systems   分布式控制系统
  IEC   国际电工委员会
  ladder logic   梯形逻辑图
  Sequential Function Charts   顺序功能图
  power supply   电源

Notes

[1] A programmable logic controller, PLC, or programmable controller is a small computer used for automation of real-world processes, such as control of machinery on factory assembly lines.

译文:可编程序逻辑控制器,PLC,或者称可编程序控制器,它是一种小型计算机,可

用于自动化生产，比如控制工厂的装配机械等。

说明：句中 used for automation of real-world processes 是过去分词短语作后置定语，修饰 computer。

［2］The input/output arrangements may be built into a simple PLC, or the PLC may have external I/O modules attached to a proprietary computer network that plugs into the PLC.

译文：输入/输出装置可以是安装在 PLC 内部，或者通过外部 I/O 模块与专用的计算机网络相连。

说明：句中 attached to a proprietary computer network 是过去分词短语作后置定语来修饰 external I/O modules。that plugs into the PLC 是定语从句，that 在从句中作主语，先行词是 computer network（专用的计算机网络）。

Exercises

Ⅰ. **Answer the following questions according to the text.**
1. How many parts do the PLCs consist of?
2. What functions can PLCs perform?
3. What is the programming language of PLC?

Ⅱ. **Decide whether the following statements are true (T) or false (F) according to the text.**
1. The instructions for the logical control sequence are stored in the data memory.
2. The I/O modules only can be built into PLCs.
3. Now, we can use structured programming languages, and logic elementary operations to program.
4. The simple ladder logic inspired from the electrical connection diagrams.
5. The programmability of a PLC is flexible.

Ⅲ. **Translate the following sentences into English.**
1. 可编程序控制器通常使用微处理器进行控制。
2. PLC 能读取限位开关、温度指示器和复杂定位系统的位置信息。
3. 最早的 PLC 采用简单的梯形逻辑图来描述其逻辑程序。
4. PLC 包含所有能直接操作大功率负载的输出功能。

Translating Skills

科技英语中一些常用的结构与表达

1）在科技文章中，要对概念或术语下定义时，常用到短语（can）be defined as，表示"是"或"被称为，定义为"的意思，其结构可以是：

术语 +（can）be defined as + 名词，例如：

Energy is usually defined as the ability to do work. 能量通常定义为做功的能力。

be defined as 的另一种常用结构是术语 +（can）be defined as + 名词 + used for doing sth / used to do sth / 定语从句或分词。例如：

Transistors are defined as the devices performing many kinds of functions in electronic equipments. 晶体管是在电子设备中执行多种功能的器件。

2）在科技文体中，如果概念或术语不需给出严格定义，只需从某种角度给予解释，此时常用的词或短语有：mean、by…we mean、in other words、be termed、be called、be named、be considered as、be known as、refer to…as、be regarded as 等。例如：

be called 被称为

This flow of electrons driven through a conductor is called an electric current. 这种被驱动通过导体的电子流称为电流。

By…we mean 所谓……是指

By memory, we mean the internal storage locations of a computer. 所谓存储器是指计算机内部的存储单元。

refer to…as… 把……称作……

We often referred to these rays as radiant matter. 我们过去常称这些射线为放射性物质。

be regarded as 把……看作

Radio waves are regarded as radiant energy. 无线电波被看作是辐射能。

be termed (= be named, be called) 被称为

The ability of a capacitor to store electrical energy is termed capacitance. 电容器存储电能的能力叫电容。

3）在科技文章中，"主语 + be + 形容词 + to 名词"的句型结构是很常见的，它用来对某一事物、概念或论点加以定论、叙述。例如：

be parallel to… 平行于……，与……平行

These lines are parallel to each other. 这些线互相平行。

be sensitive to… 对……是敏感的

The film is sensitive to light. 这种胶卷对光是敏感的。

be junior to 较年幼的，地位较低的，资历较浅的

He is junior to many other people who work here. 在这里工作的许多人当中，他资历较浅。

be applicable to 对……适应，适用于……

The formula for kinetic energy is applicable to any object that is moving. 动能公式可适用于任何运动的物体。

be corresponding to 相当于……

Zero hour Greenwich Mean Time (GMY) is corresponding to eight hours Beijing time. 格林威治标准时间零点相当于北京时间 8 点。

be particular to 是……所特有的

This mineral is particular to this region. 这种矿物是该地区所特有的。

Reading

PLC Programming

The first PLCs were programmed with a technique that was based on relay logic wiring schematics. This eliminated the need to teach the electricians, technicians and engineers how to program a computer. But this method has stuck and it is the most common technique for programming PLCs to-

100

day. An example of ladder logic can be seen in Fig. 15-2. To interpret this diagram we can imagine that the power is on the vertical line on the left hand side, we call this the hot rail. On the right hand side is the neutral rail. In the figure there are two rungs, and on each rung there are inputs or combinations of inputs (two vertical lines) and outputs (circles). If the inputs are opened or closed in the right combination the power can flow from the hot rail, through the inputs, to power the outputs, and finally to the neutral rail. An input can come from a sensor, or switch. An output will be some device outside the PLC that is switched on or off, such as lights or motors. In the top rung the contacts are normally open and normally closed, which means if input A is on and input B is off, then power will flow through the output and activate it. Any other combination of input values will result in the output X being off.

The second rung of Fig. 15-2 is more complex, there are actually multiple combinations of inputs that will result in the output Y turning on. On the left most part of the rung, power could flow through the top if C is off and D is on. Power could also (and simultaneously) flow through the bottom if both E and F are true. This would get power half way across the rung, and then if G or H is true the power will be delivered to output Y.

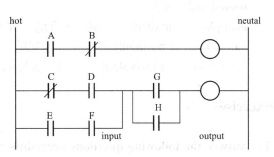

Fig. 15-2 An example of ladder logic diagram

There are other methods for programming PLCs. One of the earliest techniques involved mnemonic instructions. These instructions can be derived directly from the ladder logic diagrams and entered into the PLC through a simple programming terminal. An example of mnemonics is shown in Fig. 15-3. In this example the instructions are read one line at a time from top to bottom. The first line 0 has the instruction LDN (input load and not) for input 00001. This will examine the input to the PLC and if it is off it will remember a 1 (or true), if it is on it will remember a 0 (or false). The next line uses an LD (input load) statement to look at the input 00002. If the input is off it remembers a 0, if the input is on it remembers a 1. The AND statement recalls the last two numbers remembered and if they are both true the result is a 1, otherwise the result is a 0. The process is repeated for lines 00003 and 00004, the AND in line 5 combines the results from the last LD instructions. The OR instruction takes the two numbers now remaining and if either one is a 1 the result is

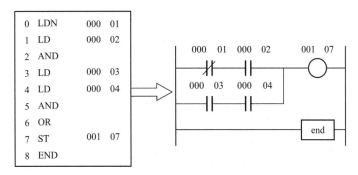

Fig. 15-3 Mnemonic & equivalent ladder logic program

a 1, otherwise the result is a 0. The last instruction is the ST (store output) that will look at the last value stored and if it is 1, the output will be turned on, if it is 0 the output will be turned off.

The ladder logic program in Fig. 15-3 is equivalent to the mnemonic program. Even if you have programmed a PLC with ladder logic, it will be converted to mnemonic form before being used by the PLC.

New Words and Phrases

    schematic   [skiː'mætik]   *adj.* 图解的，概要的   *n.* 图解视图，原理图
    hot rail   火线
    neutral rail   零线
    multiple   ['mʌltipl]   *adj.* 多样的，许多的，多重的   *n.* 并联，倍数
    mnemonic   [niː'mɔnik]   *adj.* 记忆的，助记的，记忆术的
    equivalent   [i'kwivələnt]   *adj.* 等价的，相等的，同意义的   *n.* 等价物

Exercises

I. Answer the following questions according to the text.

1. What is the first programming language of PLC?
2. How many rail in the ladder logic?
3. What is the input signal of the PLCs?
4. What is the output signal of the PLCs?
5. What is the difference between ladder logic and mnemonic?

II. Translate the following sentences into Chinese.

1. The first PLC programming eliminated the need to teach the electricians, technicians and engineers how to program a computer.

2. An output will be some device outside the PLC that is switched on or off, such as lights or motors.

3. One of the earliest techniques involved mnemonic instructions.

4. These instructions can be derived directly from the ladder logic diagrams and entered into the PLC through a simple programming terminal.

Unit 16　Electronic Measuring Instruments

Text

In general, an electronic measuring instrument is made up of the three elements shown in Fig. 16-1.

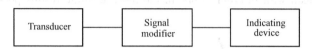

Fig. 16-1　Elements of an electronic instrument

The transducer converts a nonelectrical signal into an electrical signal; therefore, a transducer is required only if the quantity to be measured is nonelectrical (e. g. pressure) . [1]

The signal modifier is required to process the incoming electrical signal to make it suitable for application to the indicating device. [2] The signal may need to be amplified until it is of sufficient amplitude to cause any appreciable change at the indicating device. Other types of signal modifiers might be voltage dividers, to reduce the amount of signal applied to the indicating device, or wave-shaping circuits such as rectifiers, filters, or choppers.

The indicating device is generally a deflection-type meter for such general-purpose instruments as voltmeters, current meters, or ohmmeters.

Electronic measuring instruments may be used to measure current, voltage, resistance, temperature, sound level, pressure, or many other physical quantities; however, regardless of the units on the calibrated scale of the indicating meter, the pointer deflects up scale because of the flow of electrical current.

The finest instruments available may provide inaccurate results when mistreated or improperly used. There are several basic rules that, if observed, generally ensure that instruments provide acceptable measurement results. [3]

Most instruments are delicate, sensitive devices and should be treated with care. Before using an instrument one should be thoroughly familiar with its operation. The best source of information about an instrument is the operating and instructions manual, which is provided with any new instrument purchased. Electronics laboratories should have these manuals on file for easy access. If you are not thoroughly familiar with an instrument's operation, specifications, functions, and limitations, read the manual before using the instrument.

You should select an instrument to provide the degree of accuracy required. Although a high degree of accuracy and good resolution are desirable, in general, the cost of the instrument is directly related to these properties.

Once an instrument has been selected for use, it should be visually inspected for any obvious physical problems such as loose knobs, damaged case, bent pointer, loose handle, damaged test leads, and so on. If the instrument is powered by an internal battery, the condition of the battery should be checked prior to use. Many instruments have a "battery check" position for this purpose. When a battery must be replaced, make sure the proper replacement is used and that it is properly installed.

Before connecting the instrument into the circuit, make certain the function switch is set to the proper function and the range-selector switch to the proper range. If there is any question at all as to the proper range, the instrument should be set to its highest range before connecting it into the circuit; then it should be switched to lower ranges until an approximate midscale reading is obtained. There are many other considerations such as circuit loading, impedance matching, and frequency response that must be dealt with in order to obtain the most accurate results possible using test equipment.

New Words and Phrases

transducer	[trænz'dju:sə]	n.	传感器，变换器，换能器
amplify	['æmplifai]	vt.	放大，扩大，增强，详述
amplitude	['æmplitju:d]	n.	振幅，广阔，丰富，充足
appreciable	[ə'pri:ʃiəbl]	adj.	可感知的，相当可观的，可评估的
rectifier	['rektifaiə]	n.	整流器
filter	['filtə]	n.	滤波器
chopper	['tʃɔpə]	n.	斩波器
deflection	[di'flekʃən]	n.	偏向，偏差
voltmeter	['vəultmi:tə]	n.	伏特计，电压表
ohmmeter	['əum,mi:tə]	n.	电阻表，欧姆计
calibrate	['kælibreit]	vt.	标定，分度，调整，校正
pointer	['pɔintə]	n.	指针，暗示，指示器
specification	[,spesifi'keiʃ(ə)n]	n.	规格，详述，说明书
resolution	[rezə'lu:ʃ(ə)n]	n.	分辨率
knob	[nɔb]	n.	旋钮
impedance	[im'pi:dəns]	n.	阻抗
make certain			处理，应付

Technical Terms

signal modifier　信号调节器
voltage divider　分压器
waveshaping circuits　整波电路
current meter　电流表
calibrated scale　刻度盘
on file　存档
internal battery　内置电池
frequency response　频率响应

Notes

[1] A transducer is required only if the quantity to be measured is nonelectrical (e. g. pressure).

译文：只有当所测量的电量是非电信号时（例如压力）才需要传感器。

说明：在此句中，to be measured 是不定式，作后置定语，修饰 the quantity。不定式作定语常表示"将要……的"。

[2] The signal modifier is required to process the incoming electrical signal to make it suitable for application to the indicating device.

译文：用信号调节器处理输入的电信号，将其变为适用于指示装置的信号。

说明：to make it suitable 是一个带有宾语补足语的结构，it 是形式宾语，指代 the incoming electrical signal。形容词短语 suitable for application to the indicating device 是宾语补足语。

[3] There are several basic rules that, if observed, generally ensure that instruments provide acceptable measurement results.

译文：只要遵守一些基本原则，一般就可以保证仪表显示的测量数据是可以接受的。

说明：if observed 是一个省略的状语从句，全句应为 if these basic rules can be observed。当状语从句放在句子中间时，前后都需要用逗号将其与主句分开。There are several basic rules that generally ensure that…，第一个 that 引导的是定语从句，修饰 rules，第二个 that 引导的是宾语从句，作 ensure 的宾语。

Exercises

Ⅰ. Answer the following questions according to the text.

1. What is the function of the transducer?
2. What is the function of the signal modifier?
3. What is the function of the electronic measuring instruments?
4. What must to be done before using an instrument?

Ⅱ. Decide whether the following statements are true (T) or false (F) according to the text.

1. All the electronic instruments require a transducer.
2. The function of a signal modifier is to accept electrical signals.
3. To measure pressure with an electronic instrument a transducer is needed.
4. Before you use the electronic instrument for measurement, you should know clearly about how to use the device.
5. High degree of accuracy and good resolution are always the first consideration in selecting an instrument.

Ⅲ. Translate the following sentences into English.

1. 信号可能需要放大到足够的振幅以使指示装置产生明显的变化。
2. 使用仪表之前，应当完全熟悉其操作规程。
3. 应当选择其精确度能满足要求的仪表。
4. 许多仪表都有"电池检查"位置。

Ⅳ. **Translate the following paragraph.**

The company mainly engaged in electrical and electronic instruments and electronic measuring instruments manufacturing, technology development, technical services and web distribution automation business.

Translating Skills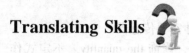

反 译 法

在英译汉的过程中常遇到这样一种情况，即原词表达的并不是其字面意义，而是其字面意义的反义，或者说是对其字面意义的否定，但这种否定又往往不出现否定词。为了符合汉语习惯，往往必须使用与原文相反的字样或句式才能确切地译出原文的含义。这种翻译方法称为"反译法"。下面分5个方面举例说明。

1. 添加否定词反译

为使译文通顺，翻译时需要使用同原文意义相反的词，可采用"否定词 + 反义词"的翻译法，在反义词前面加上否定词。例如：

Mechanical seal and ball bearing may be left assembled unless it is necessary to service them. 机械密封和滚珠轴承若不需维修，就不必拆卸。（left assembled 意为"让它装着"，此处译为"不必拆卸"更为通顺。）

There are many other sources in store. 还有多种其他能源尚未开发。

2. 删去否定词反译

为了使译文通顺，译出原文所强调的含意，翻译时需要删去原文中的否定词，再将被否定的词反译出来，使原意不变。例如：

Owing to rigidity of the spindle and bearings, the fluid bearings never lose their accuracy. 由于主轴和轴承刚性良好，流体轴承能够永久保持精度。

3. 双重反译

将句中两个否定词都反译出来，即为双重反译，可使汉语清晰、确切而严密地表达原文意义，例如：

A silicon radiation pyrometer is the only available transfer pyrometer with a stability of better than ±0.1% annually. 硅辐射高温计是唯一年不稳定性不超过 ±0.1% 范围的传热高温计。

There is no material but will deform more or less under the action of force. 在压力的作用下，任何材料或多或少都会变形。（but 是含有否定意义的关系代词，等于 that not。把 no 和 but 都译成肯定，使译文比较简明。）

4. 固定结构反译

英语中有些固定结构形似否定意为肯定，或形似肯定意为否定。译成汉语时，往往以表意为主，也可算是一种反译法。例如：

We cannot be too careful in doing experiments. 我们做实验要尽可能小心。（cannot... too 的结构是用否定的形式表示肯定意思，译为"无论怎样……也不过分"，而不是"不能太……"。）

5. 句式反译

句式反译指的是否定句和肯定句两种句式的转换翻译法。有时在翻译时必须使用与英语

相反的句式才能确切地表达原文的意思。譬如:

Don't start working before having checked the instrument thoroughly. 要对仪器彻底检查才能开始工作。

Reading

Transducers

A transducer is a device which converts the quantity being measured into an optical, mechanical, or more commonly electrical signal.[1] The energy conversion process that takes place is referred to as transduction.

1. Transducer Elements

Although there are exceptions, most transducers consist of a sensing element and a conversion or control element, as shown in the two-block diagram of Fig. 16-2.

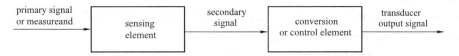

Fig. 16-2 Two-block-diagram representation of a typical transducer

For example, diaphragms, bellows, strain tubes and rings, bourdon tubes, and cantilevers are sensing elements which respond to changes in pressure or force and convert these physical quantities into a displacement. This displacement may then be used to change an electrical parameter such as voltage, resistance, capacitance, or inductance, such combination of mechanical and electrical elements form electromechanical transducing devices or transducers. Similar combinations can be made for other energy input such as thermal, photo, magnetic and chemical, giving thermoelectric, photoelectric, electromagnetic, and electrochemical transducers respectively.

2. Electrical transducers

Electrical transducers exhibit many of the ideal characteristics. In addition they offer high sensitivity as well as promoting the possible of remote indication or measurement.

Electrical transducers can be divided into two distinct groups:

(1) variable-control-parameter types, which include

* resistance
* capacitance
* inductance
* mutual-inductance types

These transducers all rely on an excitation voltage for their operation.

(2) self-generating types, which include

* electromagnetic
* thermoelectric
* photoemissive

* piezoelectric types

These all themselves produce an output voltage in response to the measurand input and their effects are reversible. For example, a piezoelectric transducer normally produces an output voltage in response to the deformation of a crystalline material[2]; however, an alternating voltage is applied across the material, the transducer exhibits the reversible effect by deforming or vibrating at the frequency of the alternating voltage.

New Words and Phrases

optical	[ˈɒptikl]	adj.	光学的；眼睛的；视觉的
transduction	[trænzˈdʌkʃən]	n.	转导；转换；换能；变频
diaphragm	[ˈdaiəfræm]	n.	隔膜；快门；[摄] 光圈；横膈膜；隔板
bellows	[ˈbeləuz]	n.	波纹管；风箱
strain	[strein]	n.	张力；拉紧 vi. 拉紧；尽力
rings	[riŋz]	n.	吊环；[电] 应力环
bourdon	[ˈbuədən]	n.	低音；带边绳；低音部音塞；钟琴的最低音
cantilever	[ˈkæntiliːvə]	n.	悬臂
displacement	[disˈpleismənt]	n.	取代；位移；[船] 排水量
electromechanical	[iˌlektrəumiˈkænikəl]	adj.	电动机械的
thermal	[ˈθəːməl]	adj.	热的；热量的；保热的 n. 上升的热气流
respectively	[riˈspektivli]	adv.	分别地，各自地；独自地
mutual	[ˈmjutʃuəl]	adj.	共同的；相互的，彼此的
photoemissive	[ˌfəutəuiˈmisiv]	adj.	光电发射的
piezo	[paiˈiːzəu]	n.	压电；压电式；压电式喷墨
measurand	[ˈmeʒərənd]	n.	被测变量；被测性能；被测情况
be referred to as			被称为
in addition			另外；此外

Notes

[1] A transducer is a device which converts the quantity being measured into an optical, mechanical, or more commonly electrical signal.

译文：传感器是一种把被测量的量转换成光的、机械的或者更普通的电信号的设备。

说明：句中"being measured"是现在分词的被动形式，表示其逻辑主语（被修饰的名词 quantity）为现在分词动作的承受者。"convert into"意思是"使转变；转化为"。

[2] A piezo-electric transducer normally produces an output voltage in response to the deformation of a crystalline material.

译文：在一般情况下，压电式传感器可根据晶体材料的变形产生一个输出电压。

说明："in response to"意思是"响应；回答；对……做出回应"在此句中翻译为："可根据"。又如：

He contributed one hundred dollars in response to my request.

108

他应我的请求而捐了 100 美元。

Exercises

I . **Answer the following questions according to the text.**
   1. What's the definition of transducer?
   2. Please give some examples of sensing elements.
   3. How does a transducer exhibit the reversible effect?

II. **Translate the following sentence into English.**
   1. 大多数传感器是由感应元件、转换器或控制元件组成的。
   2. 位移可被用来改变电参数。
   3. 传感器根据被测电量的输入生成一个输出电压。

Unit 17　Adaptive Control Systems

Text

An adaptive control system is one whose parameters are automatically adjusted to compensate for corresponding variations in the properties of the process.[1] The system is, in a word, "adapted" to the needs of the process. Naturally there must be some criteria on which to base an adaptive program. To specify a value for the controlled variable (i. e. the set point) is not enough — adaption is not required to meet this specification. Some "objective function" of the controlled variable must be specified in addition. It is this function that determines the particular form of adaption required.

The objective function for a given process may be the damping of the controlled variable. In essence, there are then two loops, one operating on the controlled variable, the other on its damping. Because damping identifies the dynamic loop gain, this system is designated as a dynamic adaptive system.

It is also possible to stipulate an objective function of the steady-state gain of the process. A control system designed to this specification is then steady-state adaptive.

There is, in practice, so little resemblance between these two systems that their classification under a single title "adaptive" has led to, much confusion.[2]

A second distinction is to be made, this not on the objective function, but rather on the mechanism through which adaption is introduced. If enough is known on a process that parameter adjustments can be related to the variables which cause its properties to change, adoption may be programmed. However, if it is necessary to base parameter manipulation upon the measured value of the objective function, adaption is effected by means of a feedback loop. This is known as a self adaptive system.

1. Dynamic Adaptive Systems

The prime function of dynamic adaptive systems is to give a control loop a consistent degree of stability. Dynamic loop gain is then the objective function of the controlled variable being regulated; its value is to be specified.

The property of the process most susceptible to change is gain. In some cases the steady state gain changes, which is usually termed nonlinearity. Other processes exhibit a variable period, which reflects upon their dynamic gain. But by whichever mechanism loop stability is affected, it can always be restored by suitable adjustment of controller gain. (This assumes that the desired degree of damping could be achieved in the first place, which rules out limit cycling.)

Many cases of variable process gain have already been cited. In generally, attempt is made to compensate for these conditions by the introduction of selected nonlinear functions into the control system. For example, the characteristic of a control valve is customarily chosen with this purpose in mind. But compensation in this way can fall short for several reasons.

1) The source of the gain variation lies in outside the loop, and hence is not identified by controller input or output.

2) The required compensation is a combined function of several variables.

3) The gain of the process varies with time.

2. The Steady-state Adaptive Problem

Where the dynamic adaptive system controls the dynamic gain of a loop, its counterpart seeks a constant steady-state process gain. This implies, of course, that the steady-state process gain is variable and that one particular value is most desirable.

Consider the example of a combustion control system whose fuel-air ratio is to be set for the highest efficiency. Excess fuel or air will both reduce efficiency. The true controlled variable is efficient, while the true manipulated variable is the fuel air ratio. The desired steady state gain in this instance is $dc/dm = 0$. The system is to be operated at the point where either an increase or decrease in ratio decreases efficiency. This is a special case of steady-state adaption known as "optimizing". A gain other than zero may reasonably be stipulated, however.

Where the value of the manipulated variable which satisfies the objective function is known relative to conditions prevailing within the process, the adaption may be easily programmed. As an example, the optimum fuel-air ratio may be known for various conditions of air flow and temperature. The control system may then be designed to adapt the ratio to air flow and temperature much in the way that the controller settings are changed, as a function of flow in the example of dynamic adaptation.[3]

New Words and Phrases

adaptation [ˌædəpˈteɪʃ(ə)n] n. 适合，适应，改编，适应性控制
designate [ˈdezɪgneɪt] vt. 指明，指定，选定，标志
confusion [kənˈfjuːʒ(ə)n] n. 混乱，混淆
prevail [priˈveil] vi. 流行，经常发生
fall short 未能满足，不能达到
parameters [pəˈræmitə] n. 参数，参量，系数
automatically [ˌɔːtəˈmætɪklɪ] adv. 自动地，机械地，不经思索地
compensate [ˈkɒmpenseɪt] vi. 补偿，赔偿，抵消 vt. 付报酬
corresponding [ˌkɒrɪˈspɒndɪŋ] adj. 通信的，一致的，相应的，v. 相配
criteria [kraɪˈtɪərɪə] n. 标准，条件（criterion 的复数）
specify [ˈspesəfai] vt. 指定，列举，详细说明
loop [luːp] vi. 打环，翻筋斗 n. 环，圈，弯曲部分 vt. 使成环
mechanism [ˈmekənɪzəm] n. 机械装置，机制，进程
manipulation [məˌnɪpjʊˈleɪʃ(ə)n] n. 操作，操纵，处理，篡改
susceptible [səˈseptəbl] adj. 易受影响的，容许……的 n. 易得病的人
counterpart [ˈkaʊntəpɑːt] n. 副本，配对物，极相似的人或物
combustion [kəmˈbʌstʃ(ə)n] n. 燃烧，氧化，骚动
ratio [ˈreɪʃɪəʊ] n. 比率，比例
optimize [ˈɒptɪmaɪz] vt. 使最优化，使完善
damping [ˈdæmpɪŋ] n. 阻尼，衰减，减幅，抑制

distinction	[dɪˈstɪŋ(k)ʃ(ə)n]	n. 差别，区别，特性，荣誉
nonlinearity	[ˌnɒnlɪniˈærəti]	n. 非线性，非线性特征
hence	[hens]	adv. 因此，今后
stipulate	[ˈstɪpjuleɪt]	vi. 规定，保证 vt. 规定 adj. 有托叶的
optimum	[ˈɒptɪməm]	adj. 最适宜的 n. 最佳效果，最适宜条件

Notes

[1] An adaptive control system is one whose parameters are automatically adjusted to compensate for corresponding variations in the properties of the process.

译文：适应性控制系统是一种能自动调整其参数以补偿与过程特性相应变化的系统。

说明：句中，whose parameters are automatically adjusted 是定语从句，不定式 to compensate for... 表示目的，译为"以弥补……"。

[2] There is, in practice, so little resemblance between these two systems that their classification under a single title "adaptive" has led to, much confusion.

译文：实际上，这两种系统之间几乎不存在相似，它们在单一名称"适应性"之下的分类已经引起了许多混淆。

说明：句中，in practice（实际上）相当于 in fact, indeed, virtually, actually；little resemblance 表示否定，so... that...（如此……以至），under a single title（在一个名称下），lead to（导致）。

[3] The control system may then be designed to adapt the ratio to air flow and temperature much in the way that the controller settings are changed, as a function of flow in the example of dynamic adaptation.

译文：用改变控制器设定值的方法来设计控制系统，使燃油-空气比适应于空气流量与温度的变化，作为动态适应系统例子中的一个流量函数。

说明：句中，不定式 to adapt 表示目的，译为"适应于、使……适应"，ratio to...（对……的比率），that 引导定语从句，修饰 way，as 这里表示作为。

Exercises

Ⅰ. Answer the following questions according to the text.

1. What is adaptive control system?

2. What is the difference between the dynamic adaptive systems and steady-state adaptive systems?

3. What is the most susceptible to change in the property of the process?

Ⅱ. Fill in the blanks of the following sentences according to the text.

1. An adaptive control system is one whose _____ are automatically adjusted to compensate for corresponding variations in the properties of the process.

2. To specify a value for the controlled variable is not enough—adaption is not required to _____ this specification.

3. It is this function that _____ the particular form of adaption required.

4. There is, in practice, so little _____ between these two systems that their classification under a single title "adaptive" has led to, much confusion.

5. Adaption is effected by means of a feedback loop. This is known as a _____.

6. The prime function of dynamic adaptive systems is to give a _____ a consistent degree of stability.

7. The characteristic of a _____ is customarily chosen with this purpose in mind.

8. Consider the example of a combustion control system whose _____ is to be set for the highest efficiency.

9. The control system may then be designed to _____ the ratio to air flow and temperature much in the way that the controller settings are changed.

Ⅲ. Translate the following sentences into Chinese or English.

1. Some "objective function" of the controlled variable must be specified in addition.

2. In some cases the steady state gain changes, which is usually termed nonlinearity.

3. However, if it is necessary to base parameter manipulation upon the measured value of the objective function, adaption is effected by means of a feedback loop.

4. 静态过程增益是变化的,而且有一个特定值是所期望的。

5. 满足目标函数的被调量数值是与该过程中主要条件相关的情况下,那么就能够容易地为适应性控制编制出程序。

Translating Skills

长难句的翻译

长句的翻译关键在于理解分析。由于连词、冠词和介词等功能词的作用,和非谓语动词及谓语动词等结构形式的存在,使得英语长句的修饰成分相当复杂,可以是单词、短语,也可以是从句,而且这些修饰成分还可以一个套一个地使用。再加上英汉句子语序上的差异,如定语和状语修饰语的位置差异、句子的逻辑安排差异等,这就使得英语句子结构复杂,长句较多,有时一段可能只有一句话,但是,只要分析得当,采用合适的方法,长句的翻译并不难。

1. 长难句分析的步骤

1) 结合具体语言环境了解全句大意。
2) 剖析全句,分清主要成分和次要成分,主句和从句。
3) 紧缩主干,确定句子的主、谓语,找出句子的基本构架,及其所属句型。
4) 疏通脉络,清理枝蔓,分清主次关系,注意上下层次,前后引衬及其逻辑关系。
5) 按汉语造句的规律,进行选词和语序的调整。
6) 反复研究译文,进行词句锤炼的润饰工夫,最后定稿。

Sample:

For a family of four, for example, it is more convenient as well as cheaper to sit comfortably at home, with almost unlimited entertainment available, than to go out in search of amusement elsewhere.

分析:

该句的骨干结构为"it is more... to do sth than to do sth else"是一个比较结构，而且是在两个不定式之间进行比较；该句中共有3个谓语结构，它们之间的关系为：it is more convenient as well as cheaper to... 为主体结构，但 it 是形式主语，真正的主语为第2个谓语结构：to sit comfortably at home，并与第3个谓语结构 to go out in search of amusement elsewhere 作比较。

综合上述翻译方法，这个句子我们可以翻译为：例如，对于一个4口之家来说，舒舒服服地在家中看电视，就能看到几乎数不清的娱乐节目，这比到外面别的地方去消遣要便宜而且方便。

2. 长难句的4种翻译方法

（1）顺译法

顺译法就是按照原文时间、方位、逻辑等顺序将长句的内容再现，例如：

Even when we turn off the beside lamp and are fast asleep, electricity is working for us, driving our refrigerators, heating our water, or keeping our rooms air-conditioned.

分析：该句子由1个主句、3个作伴随状语的现在分词以及位于句首的时间状语从句组成，共有5层意思：A. 即使在我们关掉了床头灯深深地进入梦乡时；B. 电仍在为我们工作；C. 帮我们开动电冰箱；D. 加热水；E. 或是室内空调机继续运转。上述5层意思的逻辑关系以及表达的顺序与汉语完全一致，因此，我们可以通过顺序法，译为：

即使在我们关掉了床头灯深深地进入梦乡时，电仍在为我们工作；帮我们开动电冰箱、将水加热或使室内空调机继续运转。

（2）倒译法

在翻译有些长句时，要想使译文通顺、符合汉语表达习惯，翻译时必须从后往前译，这就是倒译法。例如：

It therefore becomes more and more important that, if students are not to waste their opportunities, there will have to be much more detailed information about courses and more advice.

因此，如果要使学生充分利用他们（上大学）的机会，就得为他们提供大量关于课程的更为详尽的信息，作更多的指导。这个问题显得越来越重要。

（3）分译法

分译法是将长句拆开来，将其中从句或短语译成短句，分开来叙述，如：

Television, it is often said, keeps one informed about current events, allows one to follow the latest developments in science and politics, and offers an endless series of programs which are both instructive and entertaining.

分析：在此长句中，有一个插入语"it is often said"，3个并列的谓语结构，还有一个定语从句，这3个并列的谓语结构尽管在结构上同属于同一个句子，但都有独立的意义，因此在翻译时，可以采用分句法，按照汉语的习惯把整个句子分解成几个独立的分句，此句可译为：

人们常说，通过电视可以了解时事，掌握科学和政治的最新动态。从电视里还可以看到层出不穷、既有教育意义又有娱乐性的新节目。

（4）综合法

要求我们把各种方法综合使用，先仔细分析，或按照时间的先后，或按照逻辑顺序，顺逆结合，主次分明地对全句进行综合处理，以便把英语原文翻译成通顺的汉语句子。例如：

People were afraid to leave their houses, for although the police had been ordered to stand by in case of emergency, they were just as confused and helpless as anybody else.

尽管警察已接到命令,要做好准备以应付紧急情况,但人们不敢出门,因为警察也和其他人一样不知所措和无能为力。

Reading

Control System Components

It is well known that there are three basic components in a closed-loop control system. They are:

1) The error detector. This is a device which receives the low-power input signal and the output signal which may be of different physical natures, converts them into a common physical quantity for the purposes of subtraction, performs the subtraction, and gives out a low-power error signal of the correct physical nature to actuate the controller. The error detector will usually contain "transducers"; these are devices which convert signals of one physical form to another.

2) The controller. This is an amplifier which receives the low-power error signal, together with power from an external source. A controlled amount of power (of the correct physical nature) is then supplied to the output element.

3) The output element. It provides the load with power of the correct physical nature in accordance with the signal received from the controller.

New Words and Phrases

component   [kəm'pəunənt]   adj. 组成的,构成的, n. 成分,元件,组件
subtraction   [səb'trækʃn]   n. 减少,减法,差集
transducer   [trænz'dju:sə]   n. 传感器,变换器,换能器
amplifier   ['æmplifaiə]   n. 放大器,扩大器,扩音器
element   ['elimənt]   n. 元素,成分,要素,原理,自然环境

Exercises

Fill in the blanks of the following sentences according to the text.

1. A closed-loop control system usually contains ____ basic components.

2. The Controller is ____ which receives the low-power error signal, together with power from an external source.

3. The error detector will usually contain ____. These are devices which convert signals of one physical form to another.

4. The Output Element provides the load with power of the correct physical nature in accordance with the signal received from the ____.

Unit 18　Automatic Control System

Text

An automatic control system is a preset closed-loop control system. An automatic control system has two process variables associated with it: a controlled variable and a manipulated variable. A controlled variable is the process variable that is maintained at a specified value or within a specified range. For example, in the water tank lever control system, the storage tank level is the controlled variable. A manipulated variable is the process variable that is acted on by the control system to maintain the controlled variable at the specified value or within the specified range. [1] For the previous example, the flow rate of the water supplied to the tank is the manipulated variable.

1. Functions of Automatic Control System

In any automatic control system, the four basic functions that occur are: measurement, comparison, computation, correction. In the water tank level control system in the example above, the level transmitter measures the level within the tank. The level transmitter sends a signal representing the tank level to the level control device, where it is compared to a desired tank level. The level control device then computes how far to open the supply valve to correct any difference between actual and desired tank levels.

2. Elements of Automatic Control System

The three functional elements needed to perform the functions of an automatic control system are: a measurement element, an error detection element and a final control element. Relations among these elements and functions they perform in an automatic control system are shown in Fig. 18-1. The measuring element performs the measuring function by sending and evaluating the controlled variable. The error detection element first compares the value of the controlled variable to the desired value, and then signals an error if a deviation exists between the actual and desired values. [2] The final control element responds to the error signal by correcting the manipulated variable of the system.

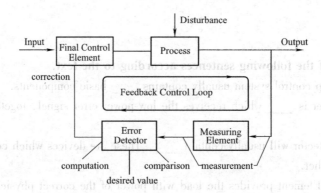

Fig. 18-1　Automatic control systems

3. Feedback Control

An automatic controller is an error-sensitive, self-correcting device. It takes a signal from the process and feeds it back into the process. Therefore, closed-loop control is referred to as feedback control.

Fig. 18-2 shows basic elements of a feedback control system as represented by a block diagram. The functional relationships among these elements are easily seen. An important factor to remember is that the block diagram represents flow paths of control signals, but does not represent flow of energy through the system or process.

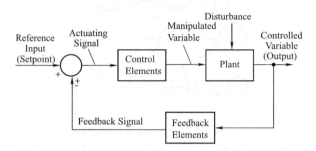

Fig. 18-2 Feedback control system block diagram

Below are several terms associated with the closed-loop block diagram. These elements are also called the "controller". The feedback elements are components needed to identify the functional relationship between the feedback signal and the controlled output. The reference point is an external signal applied to the summing point of the control system to cause the plant to produce a specified action. This signal represents the desired value of a controlled variable and is also called the "setpoint". The controlled output is the quantity or condition of the plant which is controlled. This signal represents the controlled variable. The feedback signal is sent to the summing point and algebraically added to the reference input signal to obtain the actuating signal. The actuating signal represents the control action of the control loop and is equal to the algebraic sum of the reference input signal and feedback signal. This is also called the "error signal." The manipulated variable is the variable of the process acted upon to maintain the plant output (controlled variable) at the desired value. The disturbance is an undesirable input signal that upsets the value of the controlled output of the plant.

4. Stability of Automatic Control Systems

All control modes previously described can return a process variable to a steady value following a disturbance. This characteristic iscalled "stability".

Control loops can be either stable or unstable. Instability is caused by a combination of process time lags discussed earlier and inherent time lags within a control system. This results in slow response to changes in the controlled variable. Consequently, the controlled variable will continuously cycle around the set-point value. Oscillation describes this characteristic. There are three types of oscillations that can occur in a control loop. They are decreasing amplitude, constant amplitude, and increasing amplitude. Each is shown in Fig. 18-3. Decreasing amplitude is shown in Fig. 18-3a. These oscillations decrease in amplitude and eventually stop with a control system that opposes the change in the controlled variable. This is the condition desired in an automatic control system. Constant amplitude is shown in Fig. 18-3b. Action of the controller sustains oscillations of the controlled variable. The controlled variable will never reach a stable condition; therefore, this condition is not desired. Increasing amplitude Seen in Fig. 18-3c. The control system not only sustains oscillations but also increases them. The control element has reached its full travel limits and causes the process to go out of control.

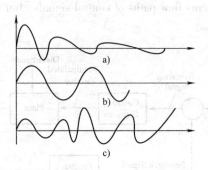

Fig. 18-3 Three types of oscillations
a) Decreasing amplitude b) Constant amplitude c) Increasing amplitude

New Words and Phrases

variable	[ˈvɛəriəbl]	adj.	易变的，多变的，n. 可变物，可变因素
manipulate	[məˈnipjuleit]	vt.	操作，操纵，使用
tank	[tæŋk]	n.	（盛液体，气体的大容器）桶、箱、罐、槽
deviation	[diːviˈeiʃ(ə)n]	n.	背离，偏离
feedback	[ˈfiːdbæk]	n.	反馈，反应
stability	[stəˈbiliti]	n.	稳定性
consequently	[ˈkɒnsikw(ə)ntli]	adv.	从而，因此
oscillation	[ˌɒsiˈleiʃn]	n.	摆动，震动
preset	[priːˈset]	adj.	预先装置的，预先调整的
specific	[spiˈsifik]	adj.	特定的，明确的，详细的，具有特效的
transmitter	[trænzˈmitə]	n.	发射机，发报机，传达人，传导物
inherent	[inˈhiərənt]	adj.	固有的，内在的，与生俱来的，遗传的
amplitude	[ˈæmplitjuːd]	n.	振幅，广阔，丰富，充足
sustain	[səˈstein]	vt.	支撑，承担，维持，忍受，供养，证实

Technical Terms

closed-loop control system　闭环控制系统
water tank lever control system　水箱水位控制系统
controlled variable　被控量
manipulated variable　控制量
error detection element　误差检测元件

118

final control element 末控制元件
decreasing amplitude 衰减振荡
constant amplitude 等幅振荡
increasing amplitude 发散振荡

Notes

[1] A manipulated variable is the process variable that is acted on by the control system to maintain the controlled variable at the specified value or within the specified range.

译文：控制量是这样一种过程变量，它是由控制系统控制，用来使被控量能保持在一个期待值或一个指定范围内。

说明：句子主干部分为 A manipulated variable is the process variable，that 所引导的定语从句是对句子主干部分的补充说明。

[2] The error detection element first compares the value of the controlled variable to the desired value, and then signals an error if a deviation exists between the actual and desired values.

译文：误差检测元件先将被控量的值与期待值作比较，如果实际值和期待值间有偏差，便给出一个信号。

说明：该句后半句"and then..."中 if 引导的是条件状语从句，强调只有在实际值和期待值有偏差时才会发出一个错误信号。

Exercises

Ⅰ. **Answer the following questions according to the text.**

1. What are the basic functions of the automatic control system?
2. Why the closed-loop control is referred to as feedback control?
3. What are the three functional elements needed to perform the functions of an automatic control system?

Ⅱ. **Fill out the blanks in the following sentences according to the text.**

1. In the water tank lever control system, ____ is the controlled variable, ____ is the manipulated variable.
2. The measuring element performs the measuring function by ____ and ____ the controlled variable.
3. There are three types of oscillations that can occur in a control loop. They are ____, ____, and ____.
4. Therefore, closed-loop control is referred to as ____.
5. Below are several terms associated with the closed-loop block diagram. These elements are also called the ____.

Ⅲ. **Translate the following sentences into Chinese or English.**

1. The level transmitter sends a signal representing the tank level to the level control device, where it is compared to a desired tank level.
2. This results in slow response to changes in the controlled variable.

3. Instability is caused by a combination of process time lags discussed earlier and inherent time lags within a control system.

4. 控制元件将会达到它们的调整极限，从而导致系统失去控制。

5. 它从控制过程中获得信号，再将信号反馈给控制过程。

Useful Information

零件数据库网站介绍与英文网站注册申请表的填写

现代的机械、电子和自动化设计早已经脱离了使用图板、圆规、直尺来绘图的手动模式，3D 设计软件已经普及。很多设计工作是先设计 3D 图样，然后再转化成 2D 工程图的形式来完成的。而现代工业的专业化分工已经很细了，从螺钉、螺母等紧固件，小的电容、电阻器、气缸电磁阀等气动部件到 PLC 乃至机器人等很多种类的工业产品都有很多知名的专业生产厂商来生产，这些厂商的 2D 和 3D 数字模型在这些公司的网站上基本都可以下载，但还有一个更方便的途径来获取这些数据，包括 2D 和 3D 图样、产品规格说明书等，这就是零件库网站（part library）。这些丰富的数据使我们无须再参照纸版说明书中的数据来重新设计已经标准化的零件甚至机器的 3D 图形来插入到我们的装配图中，从零件库网站下载的 3D 数字模型大大提升了我们的工作效率和质量。

零件库网站多数由西方发达国家的专业公司来运营，网站的缺省界面是英文的，并可以切换成几十种语言，以方便不同国家和地区的用户使用。下面介绍几个主要的零件库网站。

（1）Cadenas 零件库（PARTcommunity）

德国 CADENAS 公司推出的"零件库"分为两个产品序列，即在线版和离线版，在线版的中国（简体中文）主站已于 2010 年上线，名为 Linkable PARTcommunity，是 CADENAS 与国内的翎瑞鸿翔公司共同开发的，这个网站除囊括了 ISO／EN／DIN／GB／JIS 等主要的标准外，还包括数百家国内外厂商的产品模型，几乎涵盖了所有的机械和自动化领域，如国内用户所熟知的工业电气 ABB，魏德米勒、施耐德等等，轴承行业的 FAG／NSK／WD／NTN 等，FA 自动化的 MISUMI，气动的 SMC／FESTO／AIRTAC 亚德客等等，三维模型达数百万个规格。用户注册后，全部模型数据免费使用，其提供的众多主流 CAD 原始数据接口包括：Pro/E（Creo）、SolidWorks、UGS NX、Solid Edge、Inventor、CATIA、AutoCAD 等等，此外提供的 STEP、STL、IGES、SAT 等中间格式接口还可满足用户的不同需求。除 2D 和 3D 模型外，一些产品的规格书（specification）也可以提供。

而其离线解决方案名为 PARTsolutions，除了为 CAD 系统提供上述零部件数据资源外，还提供广泛的 ERP/PDM 接口，实现与主流 PLM（Product Life Management）系统的紧密集成，如 SAP、Intralink、Windchill、Vault、Teamcenter 和 Smarteam 等。PARTsolutions 可将其上游 ERP 系统中的物料信息与模型信息实现对接，并经 CAD 系统导入 PDM（Product Data Management），从而打通企业的整个 PLM 流程。

为适应移动通信技术的发展，微信版的 PARTcommunity 已经上线。

搜索产品时，可以在 CAD model selection 页面上点击各个厂家的 Logo（商标符号）进入其产品数据库进行查找、选型和下载，如果您知道产品的具体型号，也可以在 CAD model selection 页面的搜索栏直接搜索，待搜索成功后生成 2D 或 3D 数据，页面提供预览功能，以方便您决定是继续完善模型还是直接下载使用。

网站也提供产品的分类索引和搜索功能，产品按照工业门类进行归档。

需要说明的是，国内的基础工业起步较晚，大多数厂家被归在 Chinese manufactures 一个名目下。随着国内工业的不断进步，也有厂家被逐步单独列出。

（2）Traceparts 零件库

Traceparts 拥有全球最大且使用最广泛的数字 3D 零件库，1 亿多个 3D 数字零件模型，支持 CATIA、SolidWorks、AutoCAD、Inventor、Mechanical Desktop、Pro/E、Unigraphics 和 Solid Edge 等主流 3D 建模软件格式。所有的零件库模型开放免费下载。除了 3D 模型外，还提供很多电气或者气动符号等供设计时选用。

其标准零件库包括各个主流的工业标准，包括：GB（中国国家标准）、DIN（德国工业标准）、ISO（国际标准化组织）、ANSI（美国国家标准）、UNI（意大利国家标准）和 JIS（日本工业标准）等。

其制造商零件库包括各个门类的众多知名制造商的产品数据，如西门子、施耐德、FAG、NSK、空中客车、米其林、ABB、阿尔斯通和 SKF 等，数量达到数百家。

除提供按厂家的英文名称首字母的索引功能外，Traceparts 还提供产品的分类索引和搜索功能。如以下分类：机械系统和通用部件、液压系统和通用部件、制造工程、能量和热传导工程、电气工程、电子元件、信息技术、办公设备、新能源、图像处理、道路和车辆、飞机和航天器、原料处理设备、货物包装和调运、农业、食品技术、采矿、石油及相关设备、冶金、橡胶和塑料、建筑和建筑物、土木工程、计量和测量、文娱和体育等。

（3）其他零件库

其他类似的零件库还有 3Dpartlib 和国内的制造云等。3Dpartlib 的内容与上述两家网站类似，是 Cadenas 的合作伙伴，而制造云网站由国内公司运营，网站上面有一些课程和论坛等内容，内容在某些方面丰富一些，但标准化的 3D 资源比 Traceparts 和 PARTcommunity 要少。

需要说明的是，即使是汉语界面，在 Traceparts 和 PARTcommunity 上进行选型时，有的产品的参数只有英文说明，比如说气缸的直径为 Bore，行程为 Stroke 等，这对我们的英文水平提出了要求，当然也是我们学习技术英语的好机会，我们的科技英语水平会在工作中不知不觉地提高。

下面以 Traceparts 为例介绍一下英文网站的注册申请表的填写。下面分别是英文界面的填写内容的截图和汉语界面供对照参考。注册为网站的会员时会要求填写一些必要的联系信息，如姓名、电子邮箱、电话、通信地址、职业、工作单位名称和喜好等。除了购物网站等涉及金钱交易等内容的网站外，其他网站内容服务商一般情况下并不关注用户电子邮箱以外的信息，但电子信箱地址必须是真实的，否则当用户忘记密码时无法重新激活账户。

您可以参照图 18-4 学习英文网站注册时经常需要填写的内容。下面介绍几条下面图片中没有包含的信息或者同一信息的不同称呼，这些信息有时也被要求填写。

  Age：   年龄
  Sex：（Gender） 性别
  Marital Status： 婚姻状况
  Religious：   宗教
  Language：   语言
  Preference：   喜好
  Title：   称呼，职务

图 18-4　英文网站注册

Reading

Digital Control Systems

In a digital control system, a central processor makes the required calculations sequentially for a number of control loops. The result of a calculation may be used to drive a control valve directly or to set the set point of an analog controller. The former arrangement is known as direct digital control (DDC), and the latter as set point control (SPC). The control algorithms solved by the computer are the same in either case. But the decision whether to use a computer for DDC or SPC is sufficiently important for us to examine the benefits and limitations of each in some detail.

Fig. 18-5 shows two ways of implementing DDC. The upper loop is provided with a digital-manual station (HIC) for manual back-up. Should the computer fail, the last valve position is held and the HIC is placed in a manual mode. A light identifies the station as being in manual to attract the operator's attention. Should the operator transfer the station to manual, a logic signal is sent to the computer to report his action. When the station again returns to digital control, the computer "initializes" data stored in memory, so that automatic control can proceed bumplessly, starting at the last valve position when in manual. In the event of computer failure, all loops normally under its control must be tended by the operation. There could be as many as 100 of more, depending on the installation. Some may be sufficiently critical for manual control to be unacceptable. For these, analog backup may be chosen, as shown in the lower loop of Fig. 18-5. In the event of computer failure, an analog controller takes over regulation of the loop. However, some consideration must be given to the set point of the analog controller, or an undesirable bump may result upon transfer to analog.

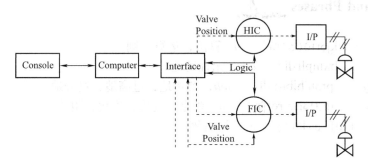

Fig. 18-5 Direct digital control can be implemented with manual backup (above) or analog backup (below)

One practice is to transfer to manual upon computer failure and leave the subsequent switch to analog control to the discretion of the operator. Another practice is to cause the analog controller's set point to track its measurement, so that there can be no deviation at the time of transfer. Alternately, the analog set point could track the digital set point, to eliminate the possibility of initiating analog control at an undesirable set point developed during a transient condition.

In the case of analog backup, two controllers have been installed (the computer plus the analog controller) although only one is in use at any given time. Consequently analog backup is prohibitively expensive except for extreme circumstances. To justify that system, the computer must be able to control the process much better than the analog controller. In actual practice, this is not likely to be the case.

If the digital computer is asked simply to duplicate the analog control function, it will not perform it as well, because the act of sampling introduces phase lag into the loop according to $Eq.$ $\Phi_\Delta = -180\Delta t/\tau_0$. If the scan period is much shorter than the loop period, this contribution may be negligible, and therefore digital control may be justified for slower loops. However, for fast loops like flow and liquid level, even a 1s sample interval may produce a noticeable deterioration in response. Yet these loops constitute the majority of loops in a typical chemical plant or petroleum refinery. To operate most of the loops at sample intervals of 1s or less loads the computer excessively; it is better reserved for complicated tasks involving extensive calculations that can be performed at less frequent intervals.

Flow and liquid level are best regulated by analog control. A flow controller can be directed by a computer, using set point control. The configuration for SPC is identical to the lower loop in Fig. 18-5 except that the computer outputs are set point rather than valve position to the FIC (flow-indicating controller). The flow controller is always controlling; a computer failure simply stops the set point from being updated. The set point may be generated by a feedback algorithm or any one of a variety of calculations based on production scheduling, feed forward control, etc. The operator may place the FIC in manual, in automatic with local set point, or in automatic with computer-driven set point. Whenever the operator returns to this last condition, the computer initializes its stored data to avoid bumping the set point the operator has introduced.

New Words and Phrases

algorithm   ['ælgərɪð(ə)m]   n. 算法，运算法则
bumplessly   ['bʌmplisli]   adv. 无扰动地
prohibitively   [prəʊ'hɪbɪtɪvli]   adv. 禁止地，过高地，过分地
deterioration   [dɪˌtɪərɪəˈreɪʃn]   n. 恶化，变质，蜕化，堕落
initialize   [ɪˈnɪʃ(ə)laɪz]   vt. 初始化

Exercises

I . **Answer the following questions according to the text.**
  1. What is DDC, and SPC?
  2. How many ways of implementing DDC?
  3. What should the operator do in the event of computer failure?
  4. In the case of analog backup, how much controllers have been installed?
  5. Which is the best way to regulate flow and liquid level?

II . **Translate the following sentences into Chinese.**
  1. The result of a calculation may be used to drive a control valve directly or to set the set point of an analog controller.
  2. Should the operator transfer the station to manual, a logic signal is sent to the computer to report his action.
  3. In the event of computer failure, an analog controller takes over regulation of the loop.
  4. A flow controller can be directed by a computer, using set point control.

Unit 19　Applications of Automatic Control

Text

Although the scope of automatic control is virtually unlimited, we will limit this discussion to examples which are commonplace in modern industry.

1. Servomechanisms

Although a servomechanism is not a control application, this device is commonplace in automatic control. A servomechanism, or "servo" for short, is a closed-loop control system in which the controlled variable is mechanical position or motion. It is designed so that the output will quickly and precisely respond to a change in the input command. Thus we may think of a servomechanism as a following device.

Another form of servomechanism in which the rate of change or velocity of the output is controlled is known as a rate or velocity servomechanism.

2. Process Control

Process control is a term applied to the control of variables in a manufacturing process. Chemical plants, oil refineries, food processing plants, blast furnaces, and steel mill are examples of production processes to which automatic control is applied. Process control is concerned with maintaining at a desired value such process variables as temperature, pressure, flow rate, liquid level, viscosity, density and composition.

Much current work in process control involves extending the use of the digital computer to provide direct digital control (DDC) of the process variables. In direct digital control the computer calculates the values of the manipulated variables directly from the values of the set points and the measurements of the process variables.[1] The decisions of the computer are applied to digital actuators in the process. Since the computer duplicates the analog controller action, these conventional controllers are no longer needed.

3. Power Generation

The electric power industry is primarily concerned with energy conversion and distribution. Large modern power plants which may exceed several hundred megawatts of generation require complex control systems to account for the interrelationship of the many variables and provide optimum power production.[2] Control of power generation may be generally regarded as an application of process control, and it is common to have as many as 100 manipulated variables under computer control.

Automatic control has also been extensively applied to the distribution of electric power. Power systems are commonly made up of a number of generating plants. As load requirements fluctuate, the generation and transmission of power is controlled to achieve minimum cost of system operation. In addition, most large power systems are interconnected with each other, and the flow of power between systems is controlled.

4. Numerical Control

There are rainy manufacturing operations such as boring, drilling, milling, and welding which must be performed with high precision on a repetitive basis. Numerical control (NC) is a system that uses predetermined instructions called a program to control a sequence of such operations.[3] The instructions to accomplish a desired operation are coded and stored on some medium such as punched paper tape, magnetic tape, or punched cards. These instructions are usually stored in the form of numbers-hence the name numerical control. The instructions identify what tool is to be used, in what way (e.g. cutting speed), and the path of the tool movement (position, direction, velocity, etc.).

5. Transportation

To provide mass transportation systems for modern urban areas, large, complex control systems are needed. Several automatic transportation systems now in operation have high-speed trains running at several-minute intervals. Automatic control is necessary to maintain a constant flow of trains and to provide comfortable acceleration and braking at station stops.

Aircraft flight control is another important application in the transportation field. This has been proven to be one of the most complex control applications due to the wide range of system parameters and the interaction between controls. Aircraft control systems are frequently adaptive in nature; that is, the operation adapts itself to the surrounding conditions. For example, since the behavior of an aircraft may differ radically at low and high altitudes the control system must be modified as a function of altitude.

Ship-steering and roll-stabilization controls are similar to flight control but generally require far higher powers and involve lower speeds of response.[4]

New Words and Phrases

scope　　　[skəup]　　n. 范围，显示器
commonplace　　['kɒmənpleɪs]　　adj. 平常的
servomechanism　　[ˌsɜːvəʊ'mekənɪzəm]　　n. 伺服机构，伺服机械
closed-loop　　['kləuzdluːp]　　n. 闭合回路；闭合环路，闭环
variable　　['veərɪəbəl]　　n. 变量；可变因素　　adj. 变量的；可变的；易变的
velocity　　[və'lɒsɪti]　　n. 速度
refinery　　[rɪ'faɪnərɪ]　　n. 精炼厂；提炼厂；冶炼厂
blast furnace　　鼓风炉，高炉
viscosity　　[vɪ'skɒsɪtɪ]　　n. 黏度，黏性
density　　['densɪtɪ]　　n. 密度
manipulate　　[mə'nɪpjʊleɪt]　　vt. 操纵；操作；巧妙地处理
measurement　　['meʒəm(ə)nt]　　n. 测量，度量；测量结果，度量制
actuator　　['æktjʊeɪtə]　　n. [自] 执行机构；制动器，执行器，传动装置
fluctuate　　['flʌktʃʊeɪt]　　vi. 波动；涨落；动摇　　vt. 使波动；使动摇
following device　　随动装置
punched paper tape　　穿孔纸带

punched card 穿孔卡片
ship-steering 船舶转向

Notes

［1］In direct digital control the computer calculates the values of the manipulated variables directly from the values of the set points and the measurements of the process variables.

译文：在直接数字控制中，计算机是根据设定点的数值和过程变量的测量值算出操纵变量值的。

说明：这是一个含有主—动—宾结构的简单句。其宾语由 of the manipulated variables 作后置定语，directly from... 引出方式状语。

［2］Large modern power plants which may exceed several hundred megawatts of generation require complex control systems to account for the interrelationship of the many variables and provide optimum power production.

译文：发电量可能超过几百兆瓦的现代化大型电厂需要复杂的控制系统来负责处理大量相互关系复杂的变量，并提供最佳的电能。

说明：which 引导定语从句，修饰 power plants，to account for... 是不定式短语作目的状语，有说明、计算之意。

［3］Numerical control (NC) is a system that uses predetermined instructions called a program to control a sequence of such operations.

译文：数字控制是这样一个系统，该系统使用的是称为程序的预定指令来控制一系列运行。

说明：that 引导定语从句，修饰 system。called a program 是过去分词短语作后置定语，修饰 instructions，不定式短语 to control a sequence of such operations 作目的状语。

［4］Ship-steering and roll-stabilization controls are similar to flight control but generally require far higher powers and involve lower speeds of response.

译文：船舶转向和颠簸稳定控制与飞行控制相似，但一般需要更高的功率和低速响应。

说明：ship-steering（船舶转向的），are similar to...（与……相似），far + 形容词比较级，表示程度加强了。又如：far better than "好多了；远远好于"。例句：I think Apple's execution of these features is far better than its competitors. 我觉得苹果对于这些功能的实现比它的竞争对手好太多了。

Exercises

Ⅰ. **Answer the following questions according to the passage.**

1. What is a servomechanism?
2. Where is process control applied in?
3. What is process control concerned?
4. What is numerical control?

Ⅱ. **Match the following phrases in column A with column B.**

　　Column A　　　　　　　　　　　Column B

1. automatic control   a. 电力分配
2. closed-loop control   b. 随动装置
3. mechanical motion   c. 过程控制
4. following device   d. 闭环控制
5. velocity servomechanism   e. 直接数字控制
6. process control   f. 机械运动
7. direct digital control (DDC)   g. 速度伺服机构
8. the distribution of electric power   h. 自动控制

III. Fill in the blanks with the proper word. Change the form if necessary.

| common | extensive | distribution |
| interconnected | minimum | generating |

    Automatic control has also been _____ applied to the _____ of electric power. Power systems are _____ made up of a number of _____ plants. As load requirements fluctuate, the generation and transmission of power is controlled to achieve _____ cost of system operation. In addition, most large power systems are ____ with each other, and the flow of power between systems is controlled.

IV. Translate the following sentences into Chinese.

1. A servomechanism is a closed-loop control system in which the controlled variable is mechanical position or motion.
2. Ship-steering controls are similar to flight control.
3. Power systems are commonly made up of a number of generating plants.
4. Automatic control has also been extensively applied to the distribution of electric power.

Practical English

怎样阅读英文招聘广告

    英语招聘广告是广告中的一种，属于非正式文体。英语招聘广告中的资格要求和工作职责通常是一条条列出来的，简洁明了、清晰明确、引人注意。

1. 注意缩略语

缩略词的使用（Abbreviations）如下所述：

    1) 首字母缩略词（Acronyms）。首字母缩略词指保留每个单词的第一个字母，而把后面的字母省略，多用于表示国家、地区和机构的专有名词、表示学位的名词和一些习惯搭配的名词。

    ① 表示机构名称，如：GE = General Electric Corporation 通用电气公司。
    ② 表示学位，如：MBA = Master of Business Administration 工商管理硕士。
    ③ 习惯搭配在一起的缩略词，如：CV = Curriculum Vitae 简历。

    2) 去尾缩略词（Words Shortened via Back-clipping）。这类缩略词把一个单词的后半部分去掉，只保留前面的 2~5 个字母。如：knowl = knowledge（知识）；loc = location（位置，场所）；除此之外，表示年份和星期的词在招聘广告中也采用去尾缩略法。

3）去元音缩略词（Abbreviations Formed by Omitting Vowels）。
这类词把单词中的元音字母去掉，只剩下辅音字母。例如：hr = hour（小时）。
4）混合缩略词（Blended Abbreviations）。这类词的构成包括以下4种情况：
保留词的开头几个字母和最后一两个字母。如：agcy = agency（经商）；appt = appointment（职务）；asst = assistant（助理）；oppty = opportunity（机会）等。
把所有元音字母去掉；同时又根据需要把中间的一两个辅音字母也去掉。如：hqtrs = headquarters 总部；stmts = statements 报告等。
只保留词的开头和末尾各一个字母。如：Jr = junior 初级；Sr = senior 高级；gd = good 好。
带'的缩略词。如：int'l = international 国际性的等。

2. 省略句的理解

（1）省略主语
招聘广告在对工作内容和性质、工作职责做出说明时通常省略主语。

（2）省略动词
省略的动词通常包括系动词"be"、实义动词"have"，助动词"do"或"does"。
1）省略系动词 be。例如，
Willing to learn and enter into new working environment.
2）省略实义动词 have。例如，
Good command of English and office.
这些情况通常出现在对资格要求的说明中。省略 have 的句子常以名词打头。
3）省略助动词 do 或 does。这一情况不太多。例如，
More than 3 years managerial experience in a sheet metal industry required.
上例中在主语后省略了 is 这个系动词。
4）省略情态动词 must 或 can。在说明工作性质和内容以及招聘要求时，通常省略情态动词，句子以动词或系动词打头。例如：Work under pressure.（work 前省略了 can）

（3）省略冠词
英语招聘广告是非正式文体，所以很多情况下冠词都被省略。Baby-sit is needed. 这里省略了不定冠词 a，表示单数。虽然 a 没有出现，读者可以从后没有"s"得出只招一名。

3. 词语褒义色彩

普通英文广告如商业广告为了推销商品会在广告中会对商品有大量的美化赞扬。而招聘广告的最终目的是让读者对某个职位产生兴趣，从而促使其采取行动—应聘，因此在英文招聘广告中经常可见广告撰写人用褒义色彩浓厚的评价性形容词来描述招聘单位概况、空缺职位的优势及未来的回报等，以便吸引读者，达到宣传的目的。例如：

A leading global company（全球领先的公司）
Interesting and rewarding position in…（有趣且回报丰厚的职位）
An attractive starting salary will be…（诱人的起薪）

4. 英文招聘广告（范例）

Marketing Assistant

Responsibility：
· Responsible for the local management of marketing and sales activities according to instructions from head office.

- Collect related information for the head office.

Requirements:
- College degree or above with good English (speaking and writing).
- Develop relationship with local media and customers.
- With basic idea of sales and marketing, related experience is preferred.
- Working experience in an international organization is a must.
- Good communication and presentation skills.

Reading

VFD system

A variable-frequency drive (VFD) is a system for controlling the rotational speed of an alternating current (AC) electric motor by controlling the frequency of the electrical power supplied to the motor.[1] A variable frequency drive is a specific type of adjustable-speed drive. Variable-frequency drives are also known as adjustable-frequency drives (AFD), variable-speed, drives (VSD), AC drives, microdrives or inverter drives. Since the voltage is varied along with frequency, these are sometimes also called VVVF (variable voltage variable frequency) drives.

A variable frequency drive system generally consists of an AC motor, a VFD controller and an operator interface (seen in Fig. 19-1).

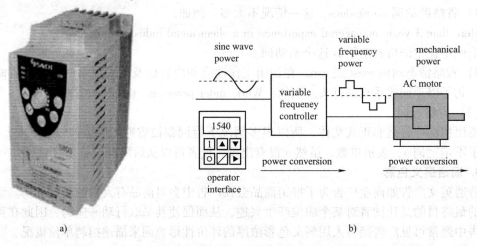

Fig. 19-1 VFD & VFD system
a) VFD controller b) VFD system

The motor used in a VFD system is usually a three-phase induction motor. Some types of single-phase motors can be used, but three-phase motors are usually preferred. Various types of synchronous motors offer advantages in some situations, but induction motors are suitable for most purposes and are generally the most economical choice. Motors that are designed for fixed-speed mains voltage operation are often used, but certain enhancements to the standard motor designs offer higher reliability and better VFD performance.

Variable frequency drive controllers are solid state electronic power conversion devices. The usual design first converts AC input power to DC intermediate power using a rectifier bridge. The DC intermediate power is then converted to quasi-sinusoidal AC power using an inverter switching circuit. [2] The rectifier is usually a three-phase diode bridge, but controlled rectifier circuits are also used (seen in Fig. 19-2).

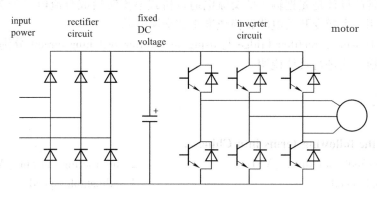

Fig. 19-2　principle VFD

The operator interface provides a means for an operator to start and stop the motor and adjust the operating speed. Additional operator control functions might include reversing and switching between manual speed adjustment and automatic control from an external process control signal. The operator interface often includes an alphanumeric display and/or indication lights and meters to provide information about the operation of the drive.

New Words and Phrases

water hammer　水锤
assembly line　流水线，装配线
handcrafting-type method　手工方式
mass production　批量生产
rate valve　速率阀
manufactured good　制造商品
rotational　[roˈteʃənl]　adj. 转动的，回转的，轮流的
synchronous　[ˈsiŋkrənəs]　adj. 同步的，同时的
rectifier　[ˈrektifaiə]　n. 整流器，改正者，矫正者
alphanumeric　[ˌælfənjuːˈmerɪk]　adj. 字母数字的

Notes

[1] A variable-frequency drive (VFD) is a system for controlling the rotational speed of an alternating current (AC) electric motor by controlling the frequency of the electrical power supplied to the motor.

译文：变频调速系统是通过控制供给电动机的频率来控制交流电动机的旋转速度的。可变频率驱动是一种调速驱动的形式。

[2] The usual design first converts AC input power to DC intermediate power using a rectifier bridge. The DC intermediate power is then converted to quasi-sinusoidal AC power using an inverter switching circuit.

译文：通常的设计是先把输入的交流电通过桥式整流转换成直流电（作为中间量），然后再把直流电用一个逆变开关电路转换成准正弦交流电源。

说明：句中 using a rectifier bridge 与 using an inverter switching circuit 分别在句子中作状语，对两种转换方式进行具体说明。

Exercises

Ⅰ. Translate the following terms into Chinese.

1. variable-frequency drive (VFD)
2. alternating current (AC)
3. rotational speed
4. adjustable-speed
5. variable-frequency
6. inverter drives
7. VVVF (variable voltage variable frequency)
8. three-phase induction motor
9. single-phase motors

Ⅱ. Translate the following sentences into English.

1. 变频调速系统是通过控制供给电动机的频率来控制交流电动机的旋转速度的。
2. 一个变频驱动系统一般由交流电动机、变频控制器和操作界面组成。
3. 变频驱动控制是固态电子技术的功率变换设备。
4. 操作者用操作界面来起动或停止电动机，调节电动机的速度。
5. 操作界面通常有图形界面显示，指示灯和仪表用来表示（系统）驱动运行的信息。

Chapter Ⅳ   Industry 4.0 & Chinese Brands Manufacturing

Unit 20   Industry 4.0 Introduction

Text

Industry 4.0's provenance lies in the powerhouse of German manufacturing. [1] However the conceptual idea has since been widely adopted by other industrial nations within the European Union, and further afield in China, India and other Asian countries. The name Industry 4.0 refers to the fourth industrial revolution, with the first three coming about through mechanization, electricity and IT.

The fourth industrial revolution, and hence the 4.0, will come about via the Internet of Things and the Internet of services becoming integrated with the manufacturing environment. [2] However, all the benefits of previous revolutions in industry came about after the fact, whereas with the fourth revolution we have a chance to proactively guide the way it transforms our world.

The vision of Industry 4.0 is that in the future, industrial businesses will build global networks to connect their machinery, factories and warehousing facilities as cyber-physical systems (CPS), which will connect and control each other intelligently by sharing information that triggers actions. [3] These cyber-physical systems will take the shape of smart factories, smart machines, smart storage facilities and smart supply chains. This will bring about improvements in the industrial processes within manufacturing as a whole, through engineering, material usage, supply chains and product lifecycle management. These are what we call the horizontal value chain, and the vision is that Industry 4.0 will deeply integrate with each stage in the horizontal value chain to provide tremendous improvements in the industrial process. [4]

At the center of this vision will be the smart factory, which will alter the way production is performed, based on smart machines but also on smart products. It will not be just cyber-physical systems such as smart machinery that will be intelligent; the products being assembled will also have embedded intelligence so that they can be identified and located at all times throughout the manufacturing process. The miniaturization of RFID tags enables products to be intelligent and to know what they are, when they were manufactured, and crucially, what their current state is and the steps required to reach their desired state. [5]

This requires that smart products know their own history and the future processes required to transform them into the complete product. This knowledge of the industrial manufacturing process is embedded within products and this will allow them to provide alternative routing in the production process. For example, the smart product will be capable of instructing the conveyor belt, which production line it should follow as it is aware of its current state, and the next production process it requires to step through to completion. Later, we will look at how that works in practice.

For now, though, we need to look at another key element in the Industry 4.0 vision, and that is the integration of the vertical manufacturing processes in the value chain. The vision held is that the embedded horizontal systems are integrated with the vertical business processes (sales, logistics and finance, among others) and associated IT systems. They will enable smart factories to control the end-to-end management of the entire manufacturing process from supply chain through to services and lifecycle management. This merging of the Operational Technology (OT) with Information Technology (IT) is not without its problems, as we have seen earlier when discussing the Industrial Internet. However, in the Industry 4.0 system, these entities will act as one.

Smart factories do not relate just to huge companies, indeed they are ideal for small-sized and medium-sized enterprises because of the flexibility that they provide. For example, control over the horizontal manufacturing process and smart products enables better decision-making and dynamic process control, as in the capability and flexibility to cater to last-minute design changes or to alter production to address a customer's preference in the products design. [6] Furthermore, this dynamic process control enables small lot sizes, which are still profitable and accommodate individual custom orders. These dynamic business and engineering processes enable new ways of creating value and innovative business models.

In summary, Industry 4.0 will require the integration of CPS in manufacturing and logistics while introducing the Internet of Things and services in the manufacturing process. This will bring new ways to create value, business models, and downstream services for small medium enterprises.

New Words and Phrases

provenance ['prɒv(ə)nəns] n. 出处，起源
powerhouse ['pauəhaus] n. 精力旺盛的人；发电所，动力室；强国
conceptual [kən'septjuəl] adj. 概念上的
afield [ə'fi:ld] adv. 在战场上；去野外；在远处；远离
hence [hens] adv. 因此；今后
integrate ['intigreit] vt. 使……完整；vi. 成为一体 adj. 整合的；完全的
cyber ['saibə] adj. 计算机（网络）的，信息技术的
whereas [weər'æz] conj. 然而；鉴于；反之
proactively [ˌprəu'æktivli] adv. 主动地
trigger ['trigə] vt. 引发，引起；触发 vi. 松开扳柄 n. 扳机
assemble [ə'semb(ə)l] vt. 集合，聚集；装配；收集 vi. 集合，聚集
embed [im'bed; em-] vt. 栽种；使嵌入，使插入；使深留脑中
miniaturization [ˌminiətʃərai'zeiʃən] n. 小型化，微型化
tag [tæg] n. 标签；vt. 尾随，紧随；连接；起浑名；添饰 vi. 紧随
alternative [ɔ:l'tɜ:nətiv] adj. 选择性的；交替的 n. 二中择一
vertical ['vɜ:tik(ə)l] adj. 垂直的，直立的；n. 垂直线，垂直面

dynamic	[daiˈnæmik]	adj. 动态的；动力学的；有活力的 n. 动态；动力
logistic	[ləˈdʒistik]	adj. 后勤学的
downstream	[ˈdaunˈstriːm]	adv. 下游地；顺流而下 adj. 下游的
accommodate	[əˈkɔmədeit]	vt. 容纳；使适应；调解 vi. 适应；调解
cater	[ˈkeitə]	vt. 投合，迎合；满足需要；提供饮食及服务
address	[əˈdres]	vt. 演说；写地址；向……致辞；处理 n. 地址；致辞

lies in   在于……
come about   发生；产生；改变方向
bring about   引起；使掉头
end-to-end   端对端；首尾相连
Operational Technology   经营技术；操作工艺
Internet of Things   物联网
cyber-physical system   信息物理系统；网宇实体系统
RFID abbr.   无线射频识别（radio frequency identification devices）
lot size   批量

Notes

[1] Industry 4.0's provenance lies in the powerhouse of German manufacturing.

译文：工业4.0起源于制造业强国德国。

说明：这是一个简单句。powerhouse的一般含义是"精力旺盛的人或发电所"，这里取其引申意义"强国"，我们将英语翻译成中文时措辞要尽量符合汉语习惯。

[2] The fourth industrial revolution, and hence the 4.0, will come about via the Internet of Things and the Internet of services becoming integrated with the manufacturing environment.

译文：第四次工业革命，以后称之为4.0，将通过物联网和互联网服务与制造环境的集成而发生。

说明：这是一个比较复杂的简单句。The fourth industrial revolution为主语，and hence the 4.0为插入语，will come about为谓语动词的将来时态，via... environment为介词短语作方式状语来修饰come about。在介词短语结构中，the Internet of Things and the Internet of services是becoming integrated with the manufacturing environment这个动名词短语的逻辑主语。因此可称为via + 动名词的复合结构作状语。

[3] The vision of Industry 4.0 is that in the future, industrial businesses will build global networks to connect their machinery, factories, and warehousing facilities as cyber-physical systems (CPS), which will connect and control each other intelligently by sharing information that triggers actions.

译文：工业4.0的远景是，将来的工业企业将建立全球网络来连接他们的机器、工厂和库房设施形成一个信息物理系统，这个系统将通过分享信息来引发动作，从而智能地连接各个部分并进行相互控制。

说明：这是一个复合句。The vision为主语中心词，is为系动词，后边的that引导表语从句。表语从句本身也是一个复合句。其中，industrial businesses为主句主语，will build global networks是动宾结构做谓语，to connect ... as cyber-physical systems是不定式结构作目的状语，而which引导的非限制性定语从句对其先行词cyber-physical systems的功能起补充说明作用。

[4] These are what we call the horizontal value chain, and the vision is that Industry 4.0 will deeply integrate with each stage in the horizontal value chain to provide tremendous improvements in the industrial process.

译文：这是我们所称的水平价值链，对水平价值链的展望是工业4.0将在这个价值链的每个阶段进行深度集成，从而使工业生产方法发生巨大的变化。

说明：本句为并列复合句。第一个分句是主系表结构的复合句，其表语是由what引导的表语从句来担当；第二个分句也是一个主系表结构的复合句，vision在第二个分句中作主句主语，is是系动词，that引导表语从句。句中的stage是阶段的意思，指代前面提到的形成value chain的engineering, material usage, supply chains, and product lifecycle management 等。to provide... 是不定式短语作目的状语。

[5] The miniaturization of RFID tags enables products to be intelligent and to know what they are, when they were manufactured, and crucially, what their current state is and the steps required to reach their desired state.

译文：缩微化的无线射频识别标签使产品变得智能并能够知道它们是什么、什么时候被生产，更为关键的是，它们目前的状况是什么和达到期望的状态还需要的步骤。

说明：本句为复合句。The miniaturization of RFID tags 为主句的主语，enables即主句谓语动词，products在主句中作宾语，to be intelligent and to know... 是两个并列的不定式短语作宾语products的补语。而to know后面跟着由what、when和后面的what引导三个宾语从句作宾语。

[6] For example, control over the horizontal manufacturing process and smart products enables better decision-making and dynamic process control, as in the capability and flexibility to cater to last-minute design changes or to alter production to address a customer's preference in the products design.

译文：例如，对水平制造过程和智能产品的控制使得我们更好地进行决策和进行动态过程控制，在能力和灵活性方面来适应最新的设计变更或者改变生产来满足顾客在产品设计方面的嗜好。

说明：这是一个复杂的简单句。control在句子中为名词做主语中心词，后面的over... 为介词短语作后置定语修饰control。enables是谓语动词，后面跟有两个并列的宾语成分better decision-making and dynamic process control；as in the capability and flexibility 后面的to cater to 和 to alter 可以理解为enable的复合谓语结构的组成部分，to address a customer's preference in the products design 则为不定式短语作目的状语。

Exercises

Ⅰ. Answer the following questions according to the text.

1. Where is the industry 4.0 concept originated from?
2. By which technologies do the first three industry revolutions come about?
3. Via what kinds of technologies will the fourth industry come about?
4. What is the horizontal value chain of industrial process?
5. What is thevertical business processes or vertical value chain?

6. What is at the center of the vision of Industry 4.0?

Ⅱ. Match the following phrases in column A with column B.

Column A   Column B
1. powerhouse   a. 标签
2. miniaturization   b. 二者选一
3. tag   c. 信息物理系统
4. alternative   d. 缩微化
5. mechanization   e. 物联网
6. Internet of Things   f. 发电所；强者
7. end-to-end   g. 机械化
8. cyber-physical system   h. 首尾相连

Ⅲ. Fill in the blanks with the proper word. Change the form if necessary.

> optimize   via   encompass   digitalized   benefit
> refer   providing   communications

Digitally connected manufacturing, often _____ to as "Industry 4.0", _____ a wide variety of technologies, ranging from 3D printing to robotics, new materials and production systems.

A move towards Industry 4.0 would _____ the private sector. Large, integrated manufacturers would find in it a way to _____ and shorten their supply chain, for example _____ flexible factories. A more _____ manufacturing would also open new market opportunities for SMEs _____ such specialized technologies as sensors, robotics, 3D printing or machine-to-machine _____.

Ⅳ. Translate the following sentences into Chinese.

1. The first industrial revolution mobilized the mechanization of production using water and steam power.

2. The Second Industrial Revolution dates from Henry Ford's introduction of the assembly line in 1913, which resulted in a huge increase in production.

3. The Third Industrial Revolution resulted from the introduction of the computer onto the factory floor in the 1970s, giving rise to the automated assembly line.

4. The vision of Industry 4.0 is for "cyber-physical production systems" in which sensor-laden "smart products" tell machines how they should be processed.

Practical English

怎样用英文写个人简历

一般英文简历主要包括个人资料、教育背景、工作经历和兴趣爱好 4 部分。

1. 个人资料（Personal Data）

（1）名字

以"李扬"为例，标准的、英文名写法是 Yang Li。双字名应将名写在前，而将姓写在后，比如，"杨晓峰"的英文写法为 Xiaofeng Yang。

（2）地址

例如，北京之后要写中国 China。邮编的标准写法是放在省市名与国名之间，即放在 China 之前。

（3）电话

电话号码前面一定加地区号，如"86-10"；电话号码隔 4 位数字加一个"-"，如"6505-2126"；地区号后的括号和号码间加空格，如（86-10）6505-2126。

2. 教育背景（Education Background）

（1）时间顺序

在英文简历中，求职者受教育情况的排列顺序是从求职者的最高教育层次写起。至于写到什么层次为止，则无具体规定，可根据个人实际情况安排。

（2）学校名和地名

学校名大写并加粗，便于招聘者迅速识别你的学历。地名右对齐，全部大写并加粗，地名后一定写中国。

（3）社会工作

担任班干部只写职务就可以了；参加过社团协会应写明职务和社团名，如果什么职务都没有，写"member of club（s）"。

（4）英语和计算机水平

个人的语言水平和计算机能力应该在此单列说明。

3. 工作经历（Work Experience）

（1）时间顺序

工作经历在排列顺序上应从当前的工作岗位写起，直至求职者的第一个工作岗位为止。也有的按技能类别分类写，这主要是为了强调个人的某种技能。

（2）公司名与地名

公司名称应大写加粗。若全称太复杂，可以写得稍微简单一些，比如 International Business Machine 在中国可简写为 IBM。

（3）公司简介

对于新公司、小公司或招聘公司不甚熟悉的某些行业的公司，略带提一下公司的简介。

（4）职务与部门

从公司名称之后的第 2 行开始写，职务与部门应加粗，每个词的第一个字母要大写，如 Manager, Finance Department。

（5）工作内容

1）要用点句。避免用大段文字，点句的长度以一行为宜，句数以 3～5 句为佳；点句以动词开始，目前的工作用一般现在时，以前的工作用过去时。

2）主要职责与主要成就。初级工作以及开创性不强的工作应把主要职责放在前面，而较高级或开创性较强的工作则应把主要成就写在前面。工作成就要数字化，精确化。在同一公司的业绩中，应秉持"重要优先"的原则。

3）工作时的培训。接受的培训可放在每个公司的后面，因为培训是公司内部的，与公司业务有关。

4. 兴趣爱好（Hobbies and Interests）

一般写两到三项强项就可以了，弱项一定不要写，不具体的爱好不写。

除了上述所说，写英文简历在用词上也要多加斟酌，比如有闯劲，用 energetic 或者 spirited 比较好，不要用 aggressive。

总之，一份好的英文简历，要目的明确、语言简练，切忌拖沓冗长，词不达意。下面是一份英文简历的示例。

Example

Personal Information：
Family Name：Xu   Name：Wenbin
Date of Birth：July 12, 1971   Birth Place：Pingdingshan
Sex：Male   Marital Status：Unmarried
Telephone：（0375）62771234   Pager：467000
E-mail：service@168.com
Work Experience：
Nov. 1998 ~ present CCIDE Inc, as a director of software development and web publishing. Organized and attended trade shows (Comdex 99).

Summer of 1997 BIT Company as a technician, designed various web sites. Designed and maintained the web site of our division independently from selecting suitable materials, content editing to designing web page by FrontPage, Photoshop and Java as well.

Education：
1992 ~ August 1996 Sept. of Automation, Henan Poly-technic Institute
Achievements & Activities：
President and Founder of the Costumer Committee
Representative in the Student Association
Computer Abilities：
Skilled in use of MS FrontPage, Win 95/NT, Project 98, Office 97, SQL software
English Skills：
Have a good command of both spoken and written English. Past CET-4
Others：
Energetic, independent and be able to work under a dynamic environment. Have coordination skills, teamwork spirit. Studious nature and dedication are my greatest strengths.

Reading

Made in China 2025 and Industrie 4.0 Cooperative Opportunities

Following the Chinese government's issuance of its *Made in China* 2025 strategy, which outlines plans to upgrade the mainland's industries, its 13th Five-Year Plan, adopted in March 2016, sets out to deepen the implementation of this strategy in the next five-year period (2016-2020). [1] While this has aroused interests as regards the development direction of Chinese industry, some industry observers have drawn parallels with Germany's Industrie 4.0 strategy, which was designed to enhance the efficiency of German industry.

It is worth noting that some have raised concerns that the two strategies may lead to intensified competition between Chinese and German industries. [2] Nevertheless, the two countries signed a memorandum of understanding to step up cooperation in the development of smart manufacturing technology in July 2015.

Indeed, the relative industrial development of the two countries, coupled with different strategic development priorities, reveal more opportunities for cooperation than competition, including in the area of industrial robots. Moreover, the different positions held by Chinese and German industries in the global supply chain also hint at further opportunities for relevant players stemming from Sino-foreign cooperation projects.

Essentially, Germany's Industrie 4.0 advocates the adoption of state-of-the-art information and communication technology in production methods as a means to further enhance industrial efficiency. This strategy is developed on the basis that Germany's strong machinery and plant manufacturing industry, its IT competences and expertise in embedded systems and automation engineering make it well placed to consolidate its position as a global leader in the manufacturing engineering industry.

Industrie 4.0 aims for intelligent production by connecting the current embedded IT system production technologies with smart processes in order to transform and upgrade industry value chains and business models. This will require Germany to enhance its research and development efforts in areas such as further integrating manufacturing systems. New industry and technical standards will be required to enable connections between the systems of different companies and devices, while data security systems will need to be upgraded to protect information and data contained in the system against misuse and unauthorized access. All of these developments are expected to enhance the efficiency and innovative capacity of German industry, while saving resources and costs.

As regards *Made in China 2025*, the focus is on innovation and quality, as well as guiding Chinese industries to move away from low value-added activities to medium and high-end manufacturing operations, rather than pursuing expansion of production capacity. The strategy is also aimed at eliminating inefficient and outdated production capacity, and helping enterprises to conduct more own-design and own-brand business. These objectives are to be facilitated by actions including the establishment of manufacturing innovation centers, strengthening intellectual property rights protection, building up new industrial standards, and facilitating the development of priority and strategic sectors.

New Words and Phrases

Industrie ['indəsti] n. 工业（德语）
cooperative [kəʊ'ɒpərətɪv] adj. 合作的；协助的；共同的
issuance ['ɪʃjuːəns] n. 发布，发行
arouse [ə'raʊz] vt. 引起；唤醒；鼓励 vi. 激发；醒来
memorandum [memə'rændəm] n. 备忘录；便笺
tap [tæp] vt. 轻敲；轻打；vi. 轻拍；n. 水龙头；轻打 vt. 采用；开发
chancellor ['tʃɑːns(ə)lə] n. 总理（德、奥等的）；（英）大臣；校长
stem [stem] n. 干；茎；血统 vt. 阻止；vi. 阻止；起源于某事物；逆行

advocate	[ˈædvəkeɪt]	vt. 提倡，主张，拥护	n. 提倡者；支持者；律师
competence	[ˈkɒmpɪt(ə)ns]	n. 能力，胜任；权限	
consolidate	[kənˈsɒlɪdeɪt]	vt. 巩固，使固定；联合	vi. 巩固，加强
misuse	[misˈjuːz]	vt. 滥用；误用；虐待	n. 滥用；误用；虐待
capacity	[kəˈpæsɪtɪ]	n. 能力；容量；资格，地位；生产力	
eliminate	[ɪˈlɪmɪneɪt]	vt. 消除；排除	
facilitate	[fəˈsɪlɪteɪt]	vt. 促进；帮助；使容易	
intellectual	[ˌɪntəˈlektʃʊəl]	adj. 智力的；聪明的；理智的	n. 知识分子
in line with	符合；与……一致		
hint at	暗示；对别人暗示……		
sets out	出发；开始；陈述；陈列		
state-of-the-art	最先进的；已经发展的；达到最高水准的		

Notes

[1] Following the Chinese government's issuance of its *Made in China 2025* strategy, which outlines plans to upgrade the main land's industries, its 13th Five-Year Plan, adopted in March 2016, sets out to deepen the implementation of this strategy in the next five-year period (2016-2020).

译文：随着中国政府对升级工业的纲要规划《中国制造 2025》战略的发布，2016 年 3 月采用的十三五计划已经开始该战略在下一个五年计划（2016～2020）的纵深部署。

说明：这是一个复合句。主句的主语是 its 13th Five-Year Plan，sets out 是主句谓语动词，后面的不定式短语 to deepen... 作宾语，Following... 是现在分词短语作时间状语，其中 which 引导非限制性定语从句，对 strategy 起补充说明作用。Adopted in March 2016 是过去分词短语作 13th Five-Year Plan 的后置定语。

[2] It is worth noting that some have raised concerns that the two strategies may lead to intensified competition between Chinese and German industries.

译文：值得一提的是已经有人提出了担心，这两个战略将使得中国和德国工业之间的竞争加剧。

说明：这是一个含有主语从句的复合句。句中第一个 that 引导主语从句 some have raised concerns 作后置主语，it 为形式主语。句中第二个 that 引导同位语从句，对 concerns 起补充说明作用。

Exercises

I. Decide whether the following statements are True (T) or False (F) according to the text.

1. There is no cooperation between China and German governments in updating their industry because of concerns that the two strategies may lead to intensified competition between Chinese and German industries.

2. The relative industrial development of China and Germany, coupled with different strategic development priorities, render more opportunities for cooperation than competition.

3. Germany has strong competences and expertise in embedded systems and automation engineering.

4. For better communication and connecting between different systems, some new industrial and technical standards need to be developed.

5. The priorities of Germany's Industrie 4.0 and China's Made in China 2025 strategies are different.

Ⅱ. **Translate the following sentences into English.**

1. 工业4.0还没有成为现实，它还只是一个理念，但它将可能带来深远的变化和影响。

2. 预测所有的环境变化是不可能的，而这些变化又是控制系统动态响应所必需的，所以可编程逻辑将变得非常重要。

3. 无论是革命性的变化还是不断演变，工业化生产将变得更加高效。

Unit 21　China's High-speed Railway

Text

China's High-speed Railway

"High speed rail travel" has been called the most revolutionary means of transport of the late 20th century and early 21st century. Today, China's high-speed railway or CRH, is the best example of such "revolutionary means of transport". In fact, CRH only began to develop in early 2004, when China issued its "Mid and Long-term Railway Network Plan", the first such development in China's history. It was in little more than six years that China's railway realized its leap-forward, which made it possible for China to head into "an era of high-speed railway".[1]

On September 28, 2010, for instance, China's homemade "Harmony-CRH380A" (Seen in Fig. 21-1), a new generation of experimental high-speed train, realized the high speed of 416.6 km an hour along the track from Shanghai to Hangzhou. This was the world's highest speed in the history of railway. China's high speed railway is now at the forefront of the world's high-speed railway, and has become a model of "China Speed" and "Made in China".

According to China's railway network development plan, by 2020, China's total mileage of railway will have almost doubled, exceeding 120 thousand kilometers, of which high-speed railway will account for over 16 thousand km.[2] By then, a grid network of China railway with four horizontal and four vertical tracks will be formed, including the Beijing-Shanghai high-speed rail line, and the Beijing-Shenzhen-Hong Kong high-speed rail line.[3] At that time, the travel time from Beijing to Shanghai will be shortened from 10 hours to less than 4 hours. The travel time from Beijing to Guan-

Fig. 21-1　Harmony-CRH380A high-speed train

gzhou will drop from 22 hours to only 6 and a half hours. From Beijing to Kunming, the travel time will also be changed from the current fastest travel time of 38 hours into no more than 8 hours. Even from the Beijing to the distant city of Urumqi, the travel time will also be reduced from 40 hours to just 11 hours.

As the American publication "Newsweek" put it: China is now engaged in a "railway revolution". These 350 kilometers per hour high-speed railway trains have made the country's vast territory "substantially smaller", and changed the country economically.

New Words and Phrases

  issue ['iʃjuː]   n. 问题；流出   vt. 发行，颁布   vi. 发行；流出
  era ['ɪərə]   n. 时代；年代；纪元
  harmony ['hɑːmənɪ]   n. 协调；和睦；融洽；调和
  forefront ['fɔːfrʌnt]   n. 最前线，最前部；活动的中心
  model ['mɒdl]   n. 典型；模范；样式   vt. 模拟；塑造   vi. 做模型；做模特儿
  mileage ['maɪlɪdʒ]   n. 英里数
  account [ə'kaʊnt]   n. [会计]账户；解释；账单   vi. 导致   vt. 认为；把……视为
  grid [grɪd]   n. 网格；格子，栅格；输电网
  publication [ˌpʌblɪ'keɪʃ(ə)n]   n. 出版；出版物；发表
  territory ['terɪt(ə)rɪ]   n. 领土，领域；范围；地域；版图
  substantially [səb'stænʃ(ə)lɪ]   adv. 实质上；大体上；充分地
  account for   对……负有责任；对……做出解释；导致；（比例）占
  Urumqi   乌鲁木齐
  leap-forward   跨越式地
  Newsweek （美国）《新闻周刊》

Notes

[1] It was in little more than six years that China's railway realized its leap-forward, which made it possible for China to head into "an era of high-speed railway".

译文：也就仅仅 6 年多，中国铁路实现了跨越式发展，中国铁路在让世人惊叹中昂首跨入了"高铁时代"。

说明：本句为复合句。主句本身是一个强调句，in little more than six years 是被强调成分。which 引导一个非限制性定语从句，对主句起补充说明作用。which 指代整个主句的内容，it 是形式宾语，真正的宾语是 for China to head into "an era of high-speed railway"。

[2] According to China's railway network development plan, by 2020, China's total mileage of railway will have almost doubled, exceeding 120 thousand kilometers, of which high-speed railway will account for over 16 thousand km.

译文：按照中国规划的宏伟蓝图，到 2020 年，中国铁路营运总里程将再增加近一倍，突破 12 万千米，其中高速铁路将超过 1.6 万千米以上。

说明：本句为复合句。主句本身是简单句，exceeding... 是现在分词短语作伴随情况的状语，of which 引导一个非限制性定语从句，对先行词 120 thousand kilometers 起附加说明作用。

[3] By then, a grid network of China railway with four horizontal and four vertical tracks will be formed, including the Beijing-Shanghai high-speed rail line, and the Beijing-Shenzhen-Hong Kong high-speed rail line.

译文：届时，中国铁路将形成包括京-沪高铁、京-深-港高铁在内的"四纵四横"的高速客运骨架网。

说明：这是一个简单句，主语中心词是 a grid network，谓语是 will be formed。主语中心词后面 of... with... 都是 network 的后置定语，另外，including... 是现在分词短语作状语，对 with... 短语起补充说明作用。

Exercises

Ⅰ. **Answer the following questions according to the text.**

1. Why is China's high-speed railway the best example of such "revolutionary means of transport"?
2. How long will China's high speed railway account for by 2020?
3. How many hours will the travel time be reduced from theBeijing to the distant city of Urumqi by high speed railway by 2020?
4. What did the American publication "Newsweek" say?

Ⅱ. **Decide whether the following are true (T) or False (F) according to the text.**

1. In early 2004 China issued its "Mid and Long-term Railway Network Plan", the first such development in China's history.
2. China's railway realized its leap-forward in little more than six years, which made it possible for China to head into "an era of high-speed railway".
3. China's high speed railway is not at the forefront of the world's high-speed railway.
4. By 2020, a grid network of China railway with four horizontal and four vertical tracks will be formed, including the Beijing-Shanghai high-speed rail line, and the Beijing-Shenzhen-Hong Kong high-speed rail line.
5. Now China is carrying out a "railway revolution".

Ⅲ. **Fill in the blanks with the proper word or phrase. Change the form if necessary.**

| employ, promote, urge, active, around |

China's new and relatively cheap expertise in this technology _____ as both an economic and diplomatic tool. Discussions on export of the technology are being held _____ the world and visiting dignitaries to the PRC _____ to use Chinese technology for their future HSR. India, for instance, is _____ seeking Chinese HSR technology, while during a meeting with Premier Li Keqiang in Beijing in mid-October, Australian Governor-General Quentin Bryce was urged _____ China's HSR technology for Australia's needs.

Practical English

如何写英文求职信

英文求职信是大学生步入社会、走向工作岗位的第一步，了解求职信的写法具有现实意义。那么，如何写好英文求职信呢？关键要处理好求职信的5个组成部分：写信动机、自我介绍、本人能力、结尾和附件。

1. 写信动机

通常求职信是针对报纸上的招聘广告而写的。若此，信中须提到何月何日的报纸，有时工作机会是从朋友或介绍所听来的，有时写信人不知某机构、公司有工作机会，毛遂自荐。不论哪一种，求职信上一定要说明写信的缘由和目的。因此首先应该表明你是在何家媒体看到招聘广告以及所要应聘的职位。参考例句：

In reply to your advertisement in today's newspaper, respectfully offer my services for the situation. 拜读今日XX报上贵公司的广告，本人特此备函应征该职位。

2. 自我介绍

写信人应述明自己的年龄或出生年月、教育背景，尤其与应征职位有关的训练或教育科目、工作经验或特殊技能。如无实际经验，略述在学类似经验亦可。参考例句：

For the past three years, I have been in the office of the XX Trading Co., where I have been an accountant. 本人曾经在XX贸易公司服务3年，担任会计工作。

3. 本人能力

这部分非常重要，因为这体现你究竟能为公司做什么，直接关系到求职的成功率。但是也要注意一定要用最少的文字表达最多的意思。参考例句：

I am able to take dictation in English and translate it rapidly into Chinese. 我会用英文做口授笔记，同时能立即将其翻译成中文。

4. 结尾

希望并请求未来的雇主允以面谈的机会，因此信中要表明可以面谈的时间。成功的求职信绝不是虎头蛇尾的，结尾一定要引起重视。参考例句：

I should be glad to have a personal interview and can furnish references if desired. 如获面试，甚感幸运。如需保证人，本人可随时提供。

5. 附件

这部分视具体情况而定，如有详细的简历或用人单位需要的材料附在求职信中时，需要注明。参考例句：

You will find enclosed an outline of my education and business training and copies of two letters of recommendation. 有关本人的学历、工作经验等项的概要，谨同函呈上两件推荐函。

Samples for Job-Application Letter

Sample 1

April 16, 2016
P. O. Box 3
Nanyang, China 000001

Dear Sir/Madame,

Your advertisement for a Network Maintenance Technician in the April 10th *Student Daily* interested me because the position that you described sounds exactly like the kind of job I am seeking.

According to the advertisement, your position requires a good college student or above in Computer Science or equivalent field and proficient in Windows NT 4.0 and Linux System. I feel that I am competent to meet the requirements. I will be graduating from Henan Polytechnic Institute this year with my B.S. degree and BEC Vantage. My studies have included courses in computer control and management and I designed a control simulation system developed with Microsoft Visual and SQL Server.

During my education, I have grasped the principles of my major subject area and gained practical skills. Not only have I passed CET-4 and BEC Vantage, but more importantly I can communicate fluently in English. My ability to write and speak English is a good standard.

I would welcome an opportunity to attend you for an interview.

Enclosed is my resume and if there is any additional information you require, please contact me (Tel: 12345678901).

Yours faithfully,
Xiao Hong

Sample 2

Dear Sir or Madam,

I am writing to apply for the position of business manager you advertised in yesterday's *Pingdingshan Daily*.

To briefly introduce myself, I am a graduate student of Pingdingshan Industrial College of Technology majoring in business administration, and expect graduation this June. Not only have I got excellent academic performance in all courses, I also possess the rich experience of assisting management staff of several renowned international companies, such as KPMG International and Microsoft. My interactive personal skills and teamwork spirit are also appropriate for this post. For further information, please refer to my attached resume.

I would be grateful if you could arrange an interview at your earliest convenience.

Yours sincerely,
Xiao Ming

Encls: 1. My resume
       2. Four recent photos

Reading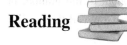

In Las Vegas: The Chinese Have Arrived

The world's largest electronics show takes place every January in Las Vegas. More than 1,100 Chinese companies, an increase from 550 last year, making up about one-third of all the 3,600 exhibitors represented at the Consumer Electronics Show (CES) in Las Vegas, Nevada, in January, 2016.[1]

"Make no mistake — the Chinese have arrived", Tim Bajarin, president of Creative Strategies, said in a recent Tech Pinions blog post. They plan to disrupt the traditional CE players as much as possible and take market share away from them fast. For top-tier Chinese companies like Huawei Technologies Co and ZTE Corp, CES remains a place to unveil new products. For new players like AvatarMind and HandScape, the show may be their foot in the door of the international market.

Another indication of China's presence at the show was that Chinese companies occupied sprawling booths at prime locations. The first booth that thousands of visitors saw when they entered the main hall was Intel Corp, the US semiconductor giant. [2] Flanking that booth were major Chinese TV makers Hisense Co Ltd and Sichuan Changhong Electric Co Ltd. Chinese companies were in every category from TVs and phones to drones and robots iPAL, a robot the height of a 3 or 4-year-old child, was developed by Nanjing-based AvatarMind as a companion for kids who "feel lonely sometimes" or "want a pal to play with".

It comes in two versions: one shaped like a boy and the other a girl. IPAL can tell stories, sing songs like Old MacDonald Had a Farm while moving its head and arms and teach Chinese and English. Children can play games with iPAL because it has a sensor to detect hand gestures. A six-inch tablet on its chest runs an Android operating system to develop applications for MyPal. The company said that soon the robot will be able to teach lessons like "Stranger Danger", which teaches children not to get too close to strangers. AvatarMind said it also is working with a Connecticut-based company that develops robotics programs for educating children with autism. [3]

Among the TV makers, Hisense, which has invested heavily in the research and development of its smart TVs, announced 22 new TV models, ranging from small 720 pixels to curved 4K Ultra HDs. Hisense's annual revenue exceeded $16 billion in 2014. Now it is number three in the global TV market, up one spot from last year, and number two in the global 4K TV market. Hisense has also been engaged in major sports sponsorships, including NASCAR, and Formula 1 racing, and it will be even more aggressive in its sponsorships.

ZTE, best known in China as a telecom-equipment maker, is also China's leading mobile device manufacturer, and has made inroads in the US with affordable technologies for smartphones. [4] It had former NBA players Larry Johnson, Larry Nance, and Adonal Foyle at its booth, along with the NBA National Championship trophy.

Lenovo, the world's No 1 personal computer maker, unveiled the world's thinnest convertible laptop, YOGA 900S, ThinkPad X1 tablet, and AirClass, a virtual classroom for corporate training.

"More and more Chinese companies will become global brands. It's hard, but I think the more successful cases there are, the easier it will become. One day we will see a boom in the number of Chinese brands," said Yang Yuanqing, CEO of Lenovo, "Companies like us have paved the road".

New Words and Phrases

represent　［ˌrepriˈzent］　vt. 代表；表现；描绘；再赠送　vi. 代表；提出异议
pinion　　［ˈpinjən］　n. [机] 小齿轮；翅膀，鸟翼　vt. 绑住，束缚；剪掉鸟翼
sprawling　［ˈsprɔːliŋ］　adj. 不规则伸展的
booth　　［buːð、buːθs］　n. 展台；电话亭

giant	[ˈdʒaiənt]	n. 巨人；伟人；adj. 巨大的；巨人般的
flank	[flæŋk]	n. 侧面 vt. 守侧面；vi. 侧面与……相接 adv. 在左右两边
pal	[pæl]	n. 朋友，伙伴；同志 vi. 结为朋友
gesture	[ˈdʒestʃə]	n. 姿态；手势 vi. 作手势；用动作示意 vt. 用动作表示
aggressive	[əˈgresiv]	adj. 侵略性的；好斗的；有进取心的；有闯劲的
tablet	[ˈtæblit]	n. 平板计算机；写字板；药片
robotics	[rəuˈbɔtiks]	n. 机器人学
sponsorship	[ˈspɔnsəʃip]	n. 赞助；发起；保证人的地位；教父母身份
virtual	[ˈvɜːtʃuəl]	adj. 虚拟的；事实上的（但未在名义上获正式承认）
revenue	[ˈrevənjuː]	n. 税收，国家的收入；收益
laptop	[ˈlæptɔp]	n. 笔记本式计算机
trophy	[ˈtrəufi]	n. 奖品；战利品 vt. 用战利品装饰 adj. 有威望的
top-tier		顶级；高层
consumer electronics		消费电子
prime location		黄金地段，场所合适的，地段显眼的

Notes

[1] More than 1,100 Chinese companies, an increase from 550 last year, making up about one-third of all the 3,600 exhibitors represented at the Consumer Electronics Show (CES) in Las Vegas, Nevada, in January, 2016.

译文：超过1100家的中国企业参加了2016年1月的内华达州拉斯维加斯消费电子展（CES），这比去年的550家增加了不少，约占3600多家参展商的1/3。

说明：这是一个较复杂的简单句。主语是More than 1,100 Chinese companies，谓语是represented，后面带一个地点状语和时间状语。an increase from 550 last year 作为插入语，对Chinese companies 起补充说明作用，making up about one-third of all the 3,600 exhibitors 为现在分词短语做后置定语，进一步补充说明Chinese companies。

[2] The first booth that thousands of visitors saw when they entered the main hall was Intel Corp, the US semiconductor giant.

译文：当成千上万的参观者进入主厅时，首先映入眼帘的是美国半导体巨头——英特尔公司的展位。

说明：本句是复合句。主句为The first booth was Intel Corp。that thousands of visitors saw 为定语从句修饰 booth。when they entered the main hall 为时间状语从句，the US semiconductor giant 是 Intel Corp 的同位语。

[3] AvatarMind said it also is working with a Connecticut-based company that develops robotics programs for educating children with autism.

译文：阿凡达机器人科技有限公司（AvatarMind）表示，它们正与位于康涅狄格州的一家公司合作，这家公司致力于开发针对自闭症儿童的机器人教育程序。

说明：这是一个复合句。主语是AvatarMind，谓语said后边是省略that的宾语从句，该宾语从句中又包括了一个that引导的定语从句，修饰company。

[4] ZTE, best known in China as a telecom-equipment maker, is also China's leading mobile

device manufacturer, and has made inroads in the US with affordable technologies for smartphones.

译文：中兴通信作为中国最有名的电信设备制造商，也是中国领先的移动设备制造商，还以用得起的智能手机技术打入美国市场。

说明：这是一个并列句。句中 ZTE 为主语，best known in China as a telecom-equipment maker 为过去分词短语作后置定语，is also …manufacturer 与 has made 是并列谓语。注意 has 前面的 and，在这里被用来加强语气，用来呼应前面描述的一些中国通信公司的行为和成绩，强调更多中国公司的进步。

Exercises

I. Decide whether the following statements are True (T) or False (F) according to the text.

1. The CES show takes place every February in Las Vegas.
2. The number of Chinese companies presented in CES increased greatly compared with that of last year.
3. Hisense shows more than 200 new models in the CES this year.
4. IPAL is a robot that has the height of a 3 or 4-year-old child.
5. Yang Yuanqing is optimistic with the trend that more Chinese companies will become international brands.

II. Translate the following sentences into English.

1. 第一届 CES 于 1967 年在纽约开幕。从那之后，很多产品在一年一度的展览上发布，其中很多改变了我们的生活。
2. 国际电子消费展是世界上正茁壮成长的消费电子技术生产商的聚集地。
3. 作为活跃的天使投资者，没有比 CES 更好的地方来发现处在消费电子方面发展早期的公司和领域，还有技术上的最新趋势。
4. 壮大你的全球品牌，增强你在国际上的存在，与技术创新者们广泛接触，建设你的网络，观看可能改变世界的最新技术。

Unit 22 Sky is the Limit for Drone Manufacturers

Text

China's state-of-the-art unmanned aerial vehicle (UAV), or drone, industry is winning the bulk of the global consumer drone market and the technology is expected to become increasingly useful in a wider range of fields in the future. [1]

The nation's largest commercial drone manufacturer, DJI, launched two kinds of consumer-level UAV, at the Consumer Electronics Show in Las Vegas in January. Its products include the Phantom 3 - 4K—a new drone in its Phantom 3 series that enables ultra high-definition video shooting—and the Inspire 1 Pro black version, a camera-equipped drone.

Last November, the Shenzhen-based UAV manufacturer launched its first agricultural drone, the MG-1, which is designed for use in spraying pesticides. [2]

Founded in 2006, DJI has about 70 percent of the world's consumer drone market. Its sales reached 400,000 units in 2014. A report in *Forbes* said the company has revenues of $500 million and a net profit of $120 million. Most of its orders came from overseas markets, with domestic sales accounting for just 10 percent of the total in 2014.

Germany's Lufthansa signed a deal with DJI about a month ago to exploit the growing market for commercial drones capable of inspecting aircraft surfaces and monitoring wind farms. [3]

According to a report issued byAnalysts International, a Beijing-based consultancy company, the domestic civil drone market is expected to grow from 3.95 billion yuan ($600 million) this year to 11.09 billion yuan in 2018. Civil drones are divided into consumer-level and industry-level equipment. Industry insiders say the current growth of the civil drone market is mainly coming from consumer-level drones, which are mainly used in the entertainment field for such things as aerial photography. They say the industry-level sector will end up being worth much more.

Ehang Inc. another Chinese drone maker, unveiled the world's first drone capable of carrying a human passenger at the Consumer Electronics Show. It is possible for the drone to be charged in as little as two hours and it is capable of flying for 23 minutes just above sea level at a speed of 100 kilometers per hour, Ehang said. The company hopes the drone can end up offering a greener alternative for commuters and will ease traffic jams.

With advancement in drone technology and policy support from the government, drones are likely to be used in future in the fields of mineral exploration, traffic administration, disaster surveillance and agriculture.

New Words and Phrases

    drone   ['drəun]   *n.* 雄蜂；嗡嗡的声音；无人驾驶飞机，遥控飞机
    phantom   ['fæntəm]   *n.* 幽灵；幻影；虚位　*adj.* 幽灵的；幻觉的
    launch   [lɔːntʃ]   *vt.* 发射（导弹、火箭等）；发起，发动

pesticide	[ˈpestisaid]	n.	杀虫剂；农药
revenue	[ˈrevənjuː]	n.	收入，收益
exploit	[ˈeksplɔit]	vt.	开发，开拓；剥削；开采
inspect	[inˈspekt]	vt.	检查；视察；检阅
charge	[tʃɑːdʒ]	vt.	使充电；使承担；指责
advancement	[ədˈvɑːnsmənt]	n.	前进，进步；提升
exploration	[ˌeksplɔːˈreiʃən]	n.	探测；探究；踏勘
surveillance	[səˈveiləns]	n.	监督；监视

state-of-the-art　最先进的；已经发展的；达到最高水准的
unmanned aerial vehicle (UAV)　无人机
Consumer Electronics Show　国际消费电子产品展销会
net profit　净利润
account for　对……负有责任；对……做出解释；导致；（比例）占
civil drone　民用无人机
mineral exploration　矿产勘探
disaster surveillance　灾情监测

Notes

[1] China's state-of-the-art unmanned aerial vehicle (UAV), or drone, industry is winning the bulk of the global consumer drone market and the technology is expected to become increasingly useful in a wider range of fields in the future.

译文：中国先进的无人机行业逐渐占据了全球无人机消费市场的大半部分，而且在未来无人机技术的应用领域将更加广泛。

说明：本句是并列句，两个分句都是简单句。句中 state-of-the-art 意思是"先进的"；the bulk of 意思是"大多数，大部分"；be expected to 意思是"预计，期望"。

[2] Last November, the Shenzhen-based UAV manufacturer launched its first agricultural drone, the MG-1, which is designed for use in spraying pesticides.

译文：去年11月，这家深圳无人机制造商发布了它的第一款农用无人机 MG-1，主要用来喷洒农药。

说明：本句是复合句。which 引导非限制性定语从句，对先行词 MG-1 作补充说明。

[3] Germany's Lufthansa signed a deal with DJI about a month ago to exploit the growing market for commercial drones capable of inspecting aircraft surfaces and monitoring wind farms.

译文：大概一个月前，德国汉莎航空公司与大疆签订了一份协议，旨在开发用于检查飞机表面和监控风力发电场的商业无人机市场。

说明：这是一个简单句。句中 to exploit... farms 是不定式短语做目的状语；capable of inspecting... wind farms 是形容词短语做后置定语，修饰前面的 commercial drones。

152

Exercises

Ⅰ. Answer the following questions according to the passage.

1. What are the two kinds of consumer-level UAV DJI launched at the Consumer Electronics Show in Las Vegas in January?
2. What is Phantom 3-4K?
3. Why did Germany's Lufthansa sign a deal with DJI about a month ago?
4. What kind of drone did Ehang Inc. unveil at the Consumer Electronics Show?

Ⅱ. Translate the following phrases into Chinese or English.

1. unmanned aerial vehicle   2. 净利润
3. high-definition video shooting   4. 民用无人机
5. mineral exploration   6. 灾情监测

Ⅲ. Fill in the blanks with the proper word. Change the form if necessary.

| compare | aircraft | dirty | drone |
| outnumber | surveillance | originate | application |

An unmanned aerial vehicle (UAV), commonly known as a _____, is an _____ without a human pilot aboard. _____ to manned aircraft, UAVs are often preferred for missions that are too "dull, _____ or dangerous" for humans. They _____ mostly in military applications, although their use is expanding in commercial, scientific, recreational, agricultural, and other _____, such as policing and _____, aerial photography, agriculture and drone racing. Civilian drones now vastly _____ military drones, with estimates of over a million sold by 2015.

Ⅳ. Translate the following sentences into chinese.

1. Many cinemas have state-of-the-art screens and sound systems, and the snacks are quite good, too.
2. The feature set between the XiaoMi Drone and DJI's Phantom 3 is almost identical—they can both be made to return home and circle around a point of interest.
3. Samsung has officially launched its mobile wallet service in China, in cooperation with local vendor UnionPay.

Practical English

面 试 技 巧

英语面试是进跨国公司或合资企业的必经之路,因此如何在英文面试中很好的阐述专业技能和表现自己的综合能力就显得尤其重要,这里介绍一些面试技巧,以便帮助大家在英文面试中脱颖而出,获得理想的工作。

1. 面试前要武装好自己

(1) Know the company 了解公司

Your knowledge of the prospective employer will contribute to the positive image you want to create. Research the company before the interview.

（2） Know the job 了解所应聘的职位

Learn everything you can about the job you're interviewing for and how your previous experience and training qualify you for this position.

（3） Know yourself 了解你自己

Review your resume before the interview to have it fresh in your mind, because it will be fresh in the mind of the person who interviews you. Better yet, have it in front of you on the table.

（4） Prepare questions of your own 准备好自己要问的问题

Employers are as interested in your questions as they are in your answers. And they'll react favorably if you ask intelligent questions about the position, the company and the industry.

（5） Be ready for any eventuality 准备好随时可能发生的情况

2. 面试过程中的要素

Keep the following in mind:

（1） Make a Good First Impression 第一印象很关键

The outcome of the interview will depend largely on the impression you make during the first five minutes.

（2） Be punctual 准时

Do whatever it takes to arrive a few minutes early. If necessary, drive to the company the night before and time yourself. Allow extra time for traffic, parking and slow elevators.

（3） Dress right 着装合体

Your clothing should be appropriate for the position you're seeking. Attire must fit well within the office and be immaculate.

（4） Shake well 握手

Show your confidence with a firm handshake. And make eye contact when you shake.

（5） Speak correct body language 利用好肢体语言

Send the right message by standing straight, moving confidently, and sitting slightly forward in your chair.

（6） Be honest 诚实

Tell the interviewer about your work skills, strengths and experience, including any volunteer work you have done.

（7） Be enthusiastic 热情

Show your clear interest in the job you are seeking and in the business. Smile and make frequent eye contact. Listen attentively and take notes.

（8） Find common ground 寻求共同点

Pictures, books, plants, etc., in the office can be conversation starters.

（9） Listening skills 倾听的技巧

Listen carefully and ask questions to probe deeper into what the interviewer is telling you.

3. 面试结束的礼貌道别也要表达准确

面试时，面试官会问及很多方面的内容，包括家庭情况，如家人对你的影响等，这些都应事先有所准备。当面试结束，礼貌的道别是很必要的，它会给面试官留下深刻的印象。

Example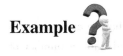

An Interview for a Mechanical Engineer
（招聘机械工程师面试范例 I：Interviewer；A：Applicant）

I: I was told by Mr. Yu that you arrived in Guangzhou this morning. You had a nice trip, didn't you?

A: Yes, I did.

I: You graduated fromHenan Polytechnic Institute?

A: Yes.

I: What department did you study in?

A: Department of Mechanical Engineering.

I: And what did you specialized in?

A: I specialized in engine designs.

I: Where are you working now?

A: I'm working at Changchun No. 1 Automobile Factory, the biggest of its kind in China.

I: As far as I know, your factory is well-known for its trucks ofJiefang brand. But do you manufacture any other vehicle?

A: Yes. We also produce buses and minibuses. And we have made advancedlimousines of Red Flag brand.

I: What is your responsibility there?

A: I'm responsible for designing engines.

I: Can you tell me something about yourachievements?

A: Of course. Years ago, I designed a kind of new diesel engine for trucks ofJiefang brand. Such a new diesel engine has decreased diesel fuel consumption by 15%. The year before last I designed a more horse powerful gasoline engine for limousines. Such a new gasoline engine has greatly increased the speed of the limousines made by our factory.

I: Wonderful. You've really made some contributions to China's automobile manufacturing. You must have received some honors or rewards?

A: Yes. I received a reward for outstanding contributions from the State Council of PRC in 2001. And I received a first-class reward for advanced designs from the People's Government of Jilin Province the year before last. In addition, I have been chosen as a "working model" in the factory over the past five years in succession.

I: OK. Now please tell me what you know about this company.

A: Your company is a Sino-Japanese joint venture which produces Honda cars. This kind of car goes fast with less gasoline consumption. So your products sell well both at home and abroad. Demand always exceeds supply.

I: Do you have any plan for the new job if we hire you as a senior mechanical design engineer with this company?

A: If I get the privilege of working at this company, I want to further improve the engine so as to raise Honda's competitiveness with such cars as Lincoln, Benz and Crown.

I: Like knows like. That is just what we want to do. It is for this reason that we are anxious to seek an aggressive and creative mechanical design engineer. I have interviewed quite a number of job seekers. But none of them are satisfactory to me. Today I've found a suitable man like you at last.

A: Thank you, sir.

I: How long can we expect you to work here?

A: If I feel I'm making progress in the work, I'll stay until the age limit.

I: Very good. We'll provide you with a yearly salary of 66,000 yuan. And we'll give you a 30-day paid vacation a year so that you can have a happy reunion with your family in Chang Chun. Do you have any particular condition you'd expect the company to take into consideration?

A: Nothing particular. But may I ask for an apartment?

I: That's out of question. We'll supply you with an apartment of two bedrooms and a living room. When can you start to work here?

A: In one month. I must go back to Changchun to hand over my work and to go through necessary procedures.

I: We'll look forward to your coming back. I wish you a nice trip, Mr. Yao.

A: Thank you, sir. See you next month.

I: Good-bye.

Reading for Celebrity Biography (Ⅱ)

Bill Gates

William (Bill) H. Gates is chairman and chief software architect of Microsoft Corporation, the worldwide leader in software, services and Internet technologies for personal and business computing.[1] Microsoft had an income of $ 25.3 billion for the financial year ending June 2001, and employs more than 40,000 people in 60 countries.

Born on October 28, 1955, Gates and his two sisters grew up in Seattle. Gates attended public elementary school and the private Lakeside school. There, he discovered his interest in software and began programming computers at age of 13.

In 1973, Gates entered Harvard University as a freshman, where he lived down the hall from Steve Ballmer, now Microsoft's chief executive officer. While at Harvard, Gates developed a version of programming language BASIC for the first microcomputer—the MITS Altair.

In his junior year, Gates left Harvard to devote his energies to Microsoft, a company he had began in 1975 with his childhood friend Paul Allen. Guided by a belief that the computer would be a valuable tool on every office desktop and in every home, they began developing software for personal computers. Gates' anticipation and his vision for personal computing have been central to the success of Microsoft and the software industry.[2]

In 1999, Gates write Business at the Speed of Thought, a book that shows how computer technology can solve business problems in fundamentally new ways. The book was published in 25 languages and is available in more than 60 countries. Business at the Speed of Thought has received wide critical applause, and was listed on the best-seller lists of the New York Times, USA Today, the Wall Street Journal and Amazon, com. Gates' previous book, The Road Ahead, published in 1995, held the No. 1 place on the New York Times' bestseller list for seven weeks.

In addition to his love of computers and software, Gates is interested in biotechnology. He is an investor in a number of other biotechnology companies. Gates also founded Corbis, which is developing one of the world's largest resources of visual information. In addition, Gates has invested with cellular telephone pioneer Craig McCaw in Teledesic, which is working on an ambitious plan to employ hundreds of low-orbit satellites to provide a worldwide two-way broadband telecommunications service.

New Words and Phrases

architect　　　['ɑːkɪtekt]　　 n. 建筑师；缔造者
corporation　　[kɔːpə'reɪʃ(ə)n]　 n. 公司；法人（团体）；社团；市政当局
elementary　　 [elɪ'ment(ə)rɪ]　　adj. 基本的；初级的；[化学] 元素的
version　　　 ['vɜːʃ(ə)n]　　n. 版本；译文；倒转术
valuable　　　['væljʊəb(ə)l]　adj. 有价值的；贵重的；可估价的　n. 贵重物品
desktop　　　 ['desktɒp]　　n. 桌面；台式机
anticipation　　 [æntɪsɪ'peɪʃ(ə)n]　 n. 希望；预感；先发制人；预支
fundamentally　 [fʌndə'mentəlɪ]　 adv. 根本地，从根本上；基础地
critical　　　　['krɪtɪk(ə)l]　　adj. 鉴定的；批评的，爱挑剔的；决定性的
applause　　　[ə'plɔːz]　　n. 欢呼，喝彩；鼓掌欢迎
bestseller　　　[best'selə(r)]　　n. 畅销书；畅销书作者；畅销商品
biotechnology　[ˌbaɪə(ʊ)tek'nɒlədʒɪ]　n. [生物] 生物技术；生物工艺学
cellular　　　 ['seljʊlə]　　adj. 细胞的；多孔的；由细胞组成的　n. 移动电话
ambitious　　 [æm'bɪʃəs]　　adj. 野心勃勃的；有雄心的；热望的；炫耀的
Seattle　　　 [si'ætl]　　n. 西雅图（美国一港市）
Harvard　　　['hɑːvəd]　　n. 哈佛大学；哈佛大学学生
Amazon　　　['æməzɔn]　　n. 亚马逊；古希腊女战士
Business at the Speed of Thought 　《未来时速》
the New York Times 　《纽约时报》
The Road Ahead 　《未来之路》
Craig McCaw　克拉格·麦考（人名）

Notes

[1] William (Bill) H. Gates is chairman and chief software architect of Microsoft Corporation, the worldwide leader in software, services and Internet technologies for personal and business computing.

译文：威廉〔比尔〕·H·盖茨是微软公司主席、首席软件设计师，也是全世界个人及商务计算机领域软件制作、服务、互联网技术的领军人物。

说明：这是一个主—系—表结构的简单句。其表语中心词为 chairman and chief software architect，of Microsoft Corporation 是后置定语；the worldwide leader 是表语中心词的同位语，in software, services…business computing 是介词短语作后置定语。

[2] Gates' anticipation and his vision for personal computing have been central to the success of Microsoft and the software industry.

译文：盖茨对个人电脑的远见卓识是微软公司和软件业成功的关键。

说明：这也是一个主—系—表结构的简单句。Gates' anticipation and his vision 是主语中心词，for personal computing 是后置定语，central（中心的，主要的）是形容词作表语，to the success…是介词短语作状语，修饰谓语 have been central。

Exercises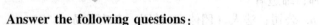

Answer the following questions:

1. Where did Gates begin to have interest in software? And when did he start programming computers?
2. What did Gates develop for the first microcomputer—the MITS Altair at Harvard University?
3. With whom did Gates start to build Microsoft?
4. What have been central to the success of Microsoft and the software industry?
5. What did Gates also found besides Microsoft?

Appendix

Appendix 1　English Communication Skills Training for Careers（职场交际技能训练）

Dialogue 1　New Guy in the Factory

(M = Michal　J = Johnson)

M: Hello, everybody. Please let me have your attention. Today we will have a new colleague, Johnson. Could you make a self-introduction first?

J: Hello, everybody. Glad to see you here. My name is Johnson. I am a newcomer here. Well, I hope we can be friends.

M: Absolutely we will. And Johnson, come on, here! Will you tell us something more about yourself, please? For example, talk about your hobby, your personality, your idea, or even your ambition if you don't mind. Everything is OK here.

J: My hobby? I like playing badminton and table tennis. Well, I just hope to do something here.

M: That's all right! Then have you got any expectation of us, Johnson?

J: Maybe I will have difficulty in my future work, so I expect you can help me. That's all, thank you.

参考译文：

工厂新同事

迈克尔：大家好。请注意一下，今天我们来了一位新同事，约翰逊。请你为我们做个自我介绍好吗？

约翰逊：大家好，很高兴见到你们。我叫约翰逊，是新来的。希望我们能成为好朋友。

迈克尔：当然会的。约翰逊，到这来！如果你不介意的话，能给大家详细地再介绍下自己吗？比如说你的业余爱好、个性、想法，甚至是你的抱负。总之，你可以讲关于你的一切。

约翰逊：我的爱好？我平时喜欢打羽毛球和乒乓球。对了，我希望在这里好好工作。

迈克尔：好的，约翰逊，你对我们大家有什么期待吗？

约翰逊：可能未来的工作还需要大家的帮助，在此我先谢谢大家了！

Dialogue 2　Introduction of the Car

(A = Adans　B = Brown)

A: What kind of car is best for me?

B: For you, I would like to recommend this one here.

A: The compact? Why do you think it's better than others?

B: This automobile just meets your needs because not only is it compact for ease in city driving, but also, it has many safety features. Finally, it can easily drive your family to enjoy your vacation during holidays.

A: That's true. Thank you for recommendation.

参考译文:

汽 车 导 购

亚当斯：请问哪种车适合我？

布朗：对你来说，我认为这辆车挺适合你的。

亚当斯：小型车？它哪些方面优于别的车？

布朗：这种车能够很好地满足你的需求。因为首先它是小型车，可以很方便地在城市里行驶。其次，它有很多安全特性。还有，它可以很方便地载着你全家去度假。

亚当斯：是的，谢谢你的介绍。

Dialogue 3　Finding out New Problem

(L = Lucas　B = Bill)

L: Hello, sir. What's the matter with this machine?

B: Maybe the spark plug or the starter can't work.

L: Is it serious?

B: No, I think I get it.

L: What?

B: The carburetor broke.

参考译文:

发现新问题

卢卡斯：先生，这台机器怎么了？

比尔：可能是火花塞或者是启动器坏了。

卢卡斯：很严重吗？

比尔：不，我想我找到答案了。

卢卡斯：什么问题？

比尔：是汽化器出问题了。

Dialogue 4　Solving Problem

(A = Annie　B = Brown)

A: What's up, Mr. Brown?

B: You know I am a green hand as a dentist, so I don't know how to use this new machine.

A: Don't lose heart. This machine has abundant countenances, and it can be as a model for young dentists according to the robot's reaction. I think you can use it smoothly.

B: Thanks, I will try my best.

参考译文：

解决问题

安妮：怎么啦，布朗先生？

布朗：你知道我是一个牙医新手，我不知道如何使用这台新机器。

安妮：别灰心。这台机器有丰富的面部表情，它能作为年轻牙医的模特。根据机器人的反应，年轻牙医可以在练习中取得进步。我想你会很好地使用这台机器的。

布朗：谢谢，我会尽我所能的。

Dialogue 5 Preparing Product Presentation

(S = Smith　B = Brooks)

S: Good morning!

B: Good morning!

S: Have you prepared the guidebook, which includes the introduction and schedules of the fair?

B: Yes, I have also prepared a fair memo.

S: Very good. By the way, will you help me to type the letters and mail them? And please phone Mr. Smith in New York and confirm his flight.

B: I will do that right after I finish typing these letters.

参考译文：

产品发布会

史密斯：早上好！

布鲁克斯：早上好！

史密斯：那本包括有会展介绍和会展时间表的小册子准备好没有？

布鲁克斯：我已经准备好了。我还准备了会展备忘录

史密斯：那很好。顺便问一下，你能帮我把这些信件打印出来，并且邮寄出去吗？另外请打电话确认一下位于纽约的史密斯先生的航班。

布鲁克斯：好的，打印完这些信我马上就去做。

Dialogue 6 Products Presentation Completion

(A = Angel　B = Bruce)

A: Hello, this is investigation center of our products. We want to get some information about our product presentation.

B: Ok, we have made some achievements. We have established business relationships with more than 40 target clients.

A: Very good. Have you completed the product presentation for this activity?

B: Oh, yes, it's all right. All the staff are very professional and they ensured our product presentation smoothly.

A: Thank you.

参考译文：

产品发布会完成

安琪：你好，我们是公司产品的调研部门，想征求您对公司产品发布会情况的意见和

建议。

布鲁斯：你好，我们取得了显著的成效。我们已经跟40多个目标客户建立了商业关系。

安琪：这很好，请问你们针对此次活动的产品发布会结束了吗？

布鲁斯：哦，是的，一切都很好。参与此次展会的全体员工都非常专业，他们也确保了我们产品发布会的顺利进行。

安琪：谢谢。

Dialogue 7　Talking About a New Order

（H = Hawk　B = Barly）

H: Now we are especially interested in your products.

B: We are glad to hear that. May I know what items you are particularly interested in?

H: Machine parts. Will you please quote us a price?

B: All right. Here are our F. O. B. price lists. All the prices in the lists are subject to our confirmation.

H: I'm sorry to say that the price you quote is too high. It would be very difficult for us to push any sales if we buy it at this price.

B: Ok! In view of the first time we cooperate, and for our long-term cooperation in future, I will give you a special discount.

H: It is very kind of you.

参考译文：

讨论新订单

霍克：目前我们对你方产品尤其感兴趣。

巴里：听你这样说我们真高兴。我能知道你具体对我方哪些商品感兴趣吗？

霍克：机器零件。你能开个价吗？

巴里：没问题。这是我们的离岸价单，里面所有的价格都以我方确认为准。

霍克：很遗憾地说，你方报价太高，如按此价格，我们将不会达成任何交易。

巴里：好的，鉴于我们第一次合作，同时为了以后能长期合作，我方将给贵方一个特殊的折扣。

霍克：那真是太好了！

Dialogue 8　Negotiation About the New Amplifier

（A = Arrow　B = Bob）

A: Mr. Mogan said he would accept 2000 units of our new amplifiers if we lower our price to 15 yuan.

B: He is driving a hard bargain. Will we make any money on this order if we agree to his price?

A: Definitely. Not a large profit, though.

B: OK. Make it to that. Go to fill him in on the good news.

参考译文：

新款扩音器的协商

艾柔：摩根先生说如果我们把报价降到15元他可以订购我们2000件新款扩音器。

鲍勃：他真会杀价。如果我们接受他的价格，还有利润吗？
艾柔：绝对有，只是不多而已。
鲍勃：那好吧，就按这个价格走，去告诉他这个好消息吧。

Dialogue 9　Promotion Plan Discussing

(A = Ally　B = Billy)

A: Good morning! It's time to discuss how we should promote our new robots.

B: I think we definitely need to choose Internet as a means of marketing. It's the current trend to do so.

A: All right! Can you say it in exact words?

B: We can create a promotion page on a website, and we should do it in an attractive manner.

A: Thank you! Let's make it!

参考译文：

讨论推销计划

阿里：早上好！我想现在我们可以讨论下公司新的机器人的推销问题了。
比利：我想我们应该选择因特网作为我们主要的营销工具，这是当前的趋势。
阿里：好的，你能说得再清楚一点儿吗？
比利：我们可以在一个网站上建一个推销的网页，并且尽量把网页做得更加吸引人。
阿里：好的，就这样定了！

Dialogue 10　Robot Promotion

(M = Michal　B = Barley)

M: How are you! I heard that our new robots are selling like hot cakes.

B: Fine. They are selling like crazy! Please look at these promotion figures.

M: Great! All of you are a great asset to our company. I'm so proud of you all. By the way, do you think our sales will keep on going like this?

B: Yes, the quality of our new products is superior to others and our price is very competitive, so our sales will be promising.

参考译文：

机器人销售

迈克尔：大家好！听说我们的新型机器人卖得非常好！
巴里：是的，何止是非常好，简直是卖疯了！快看一下这些销售数据吧！
迈克尔：好的。大家都是公司的重要人才。我为你们感到骄傲。对了，大家认为我们的销售情况会一直这样吗？
巴里：会的。我们的产品质量远优于别人，价格也极具竞争力，所以我们的销售会一直兴旺。

Dialogue 11　Cooperation Intention

(D = Darwin　C = Campbell)

D: We're interested in your products, and would like to talk something about that.

C: We worked on vacuum cleaners only for two years, but we are in a position to place large

163

orders with competitive suppliers. Every year, we export a lot of our products to European countries, but yours seems quite new to us.

D: Glad to hear that. I think we can have a large order with you if your products can meet our demand.

C: That's great! Hope to have a new chance with you.

参考译文:

合 作 意 向

达尔文：我们对贵公司的产品很感兴趣，并且想要和你谈谈。

坎贝尔：我们公司主营吸尘器虽然只有两年时间，但和同行业公司相比，我们每年的订货量很大。我们每年都要向欧洲各国出口大量的产品。不过你们公司对于我们来说，还是新客户。

达尔文：听到这些非常高兴！如果贵公司的产品能够符合我们的要求，我们会给你们一个大的订单。

坎贝尔：那太好了，希望和你们有新的机会合作。

Dialogue 12 Signing the Contract

(C = Carter   B = Ben)

C: Any questions?

B: Yes. There are a few points which I'd like to bring up. First, the packing; second, about the terms of payment; and the third point is about arbitration.

C: Now, the first point is about packing. We agree to a different packing for the bed truck. Second, we agree to pay by L/C. As for arbitration, personally I hope you'll accept this clause.

B: That's Perfect.

参考译文:

签 合 同

卡特：还有什么问题吗？

本：是，有几点我想提一下。首先，是包装；第二点是付款的条件；第三点有关仲裁。

卡特：首先，第一，有关包装，我们同意汽车底座换用不同的包装。第二，我们同意用信用证付款。至于仲裁，从我个人来讲，我希望你们接受这一条款。

本：那太好了。

Dialogue 13 Inspiring Your Staff

(A = Henry   B = Beth)

A: Beth, you have done a good job. I just come back from Sunshine Limited, and they decided to sign a one-year contract with us.

B: Oh, I'm glad that I listened to your advice first.

A: Excellent! I told you so. Where there is a will, there is a way.

B: Thanks for your advice.

A: That's all right! I'm pleased with your ability and strength.

参考译文：

鼓励你的员工

亨利：贝斯，你做得很好。我刚从阳光有限公司回来，他们同意跟我们签署一年的合约。

贝斯：哦，那太好了！幸亏当初听了您的建议。

亨利：是你做得漂亮，我说过，"有志者，事竟成。"

贝斯：那还是很感谢您的建议。

亨利：不客气，你精明能干，我很满意。

Dialogue 14　Setting a Target

（A = Anna　P = Peter）

A：Good morning, everyone! Today we shall discuss about our annual sales target of our department. Let's get into the subject first.

P：OK. I will post our sales target on the Internet after meeting.

A：I did some analysis on all the factors and I hope that sales of this year should be raised. Is there any problems?

P：No. We will try our best to achieve the goal!

参考译文：

设 定 目 标

安娜：各位，早安！今天我们要讨论本部门的年度销售目标。我们直接进入主题吧！

皮特：好的。会后我会把它放到网络上的。

安娜：我对公司各方面的因素都做了分析。我希望本年度的销售额能够增加。大家说有问题吗？

皮特：没问题。我们会尽最大努力达成目标。

Dialogue 15　Raising Salaries

（K = Kevin　J = Joseph）

K：Joseph, I would like to have a talk with you. Please have a seat.

J：Thank you.

K：After our discussion, I decided to give you a salary increase from next month, you will get a 1000 yuan raise to your salary.

J：Thank you very much. I understand that you have your consideration and I am satisfied with the raise.

K：I think you deserve this.

J：I will continue to do my best and contribute to the company.

参考译文：

加　薪

凯文：约瑟夫，我想和你谈谈，请坐！

约瑟夫：谢谢。

凯文：经过我们讨论，我决定上调你的薪资。从下个月开始，你的薪资将上调1000元。

约瑟夫：谢谢。我明白您有您的考虑，我对加薪感到满意。

凯文：我认为这是你应得的。
约瑟夫：我会继续尽最大的努力，为公司多做贡献。

Dialogue 16 Achieving the Goal

(G = Green   J = James)

G: Good evening, everyone! First, I would like to thank all of you for devoting yourselves to the job and achieving our company's goal.

J: Hooray!

G: Please raise your cup and toast for our excellent achievement in the last year.

J: Wow!

G: Thank you very much. At last, I have good news for you. All of you will be rewarded with an extra bonus.

J: I couldn't agree more. Yes, we are happy with the result although we worked like hell for the past few months.

参考译文：

实现目标

格林：大家晚上好！首先，我要感谢各位为实现公司的目标所付出的辛勤工作。

詹姆斯：万岁！

格林：请举起你们手中的酒杯并为我们过去一年所取得的优异成绩而干杯！

詹姆斯：哇！

格林：非常感谢大家。最后，我还有一个好消息要告诉大家：公司决定给你们额外的奖金作为鼓励。

詹姆斯：我非常同意。是的，虽然过去几个月我们拼了命地工作，但我们为取得的结果感到高兴。

Appendix 2　Reference Translations and Keys to Parts of Exercises

（参考译文与部分习题答案）

第1章　电力电子基础

第1单元　电路与电气元件

课文　　　　　　　　　　　　　电　路

1. 电流

电流是指带电粒子的流动。在铜线里面，电流由叫作电子的微小负电荷粒子运载。在电流开始流动之前，电子是随机方向漂移的。当电流流动时，电子开始朝一个相同的方向移动。电流的大小取决于每秒经过的电子数量多少。

电流由符号 I 表示，度量单位为安［培］（A）。1A 表示每秒有 6.24×10^{18} 个电子穿过导线的任何一个位置。这表示每秒有比 6 百万千亿个还多的电子穿过。

在电路中，电流经常用毫安（mA）表示，即千分之一安。

2. 电压

在手电筒电路中，是什么让电流流动呢？答案是电池单元提供一个"推力"，让电流在电路中环行流动。

每个电池单元都提供一个推力，这叫作它的电势差，或者叫电压。电压由符号 U 表示，度量单位是伏［特］（V）。

通常，每个电池单元提供 1.5V 的电压。2 个电池单元串联就提供 3V 电压，3 个电池单元串联就提供 4.5V 电压，如图 1-1 所示。

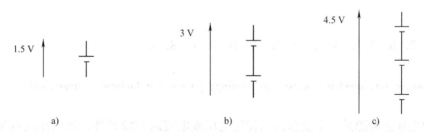

图 1-1　电池单元串联
a）1 个电池单元　b）2 个电池单元串联　c）3 个电池单元串联

3. 电池单元串联

哪种电池排列可以让灯泡最亮？灯泡是设计成在特定电压下工作的，但是，当其他条件相同时，电压越高，灯泡就越亮。

严格来说，一个电池包括两个或两个以上的电池单元。它们可以串联，就像用在手电筒电路中那样，但是它们也可以并联，如图 1-2 所示。

167

4. 电池单元并联

单个电池单元可以长时间提供小电流，或者短时间提供大电流。把电池单元串联可以提高电压，但不会影响电池单元的使用寿命。另一方面，如果电池单元是并联，电压就保持在1.5V，但是，电池寿命就延长一倍。

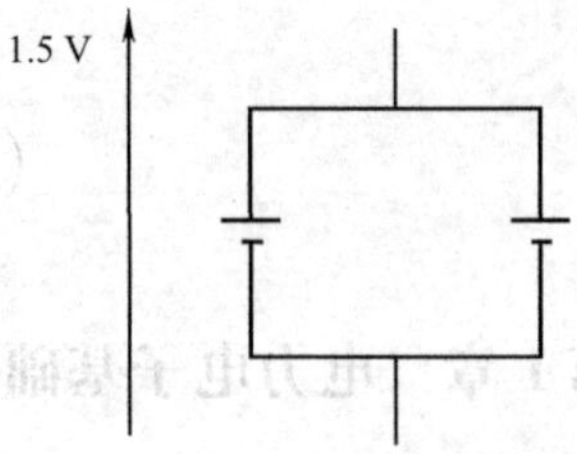

一个手电筒灯泡需要300mA电流，一个C型碱性电池可以供电超过20个小时，直到耗尽为止。

图1-2 电池单元并联

5. 电阻

手电筒电路的某些部分会限制或阻碍电流的流动。电路大部分由厚金属导体组成，使得电流可以容易流动。这些部分都是低电阻，包括弹簧，开关簧片和灯泡连接点。但是灯泡的灯丝，是由非常细的金属丝构成，它比电路的其他部分更不容易传导电流，且具有更高的电阻。

灯丝的电阻 R 的度量单位是欧［姆］（Ω）。如果电池的电压是3V（两个C型的电池单元串联），流经灯泡的电流是300mA，或者0.3A，那么灯丝的电阻是多少？

计算如下：

$$R = \frac{U}{I} = \frac{3}{0.3} = 10\Omega$$

其中，R是电阻，U是电压，I是电流。这里，10Ω是灯丝通电时的电阻。

在电路中电阻值可以是几欧（Ω），几千欧（kΩ）甚至几兆欧（MΩ）。设计成具有特定电阻值的电子元件称为电阻器。

Keys to Exercises of the Text

Ⅰ.

1. An electric current is a flow of charged particles.
2. The size of the current depends on the number of electrons passing per second.
3. The cells provide a "push" which makes the current flow round the circuit.
4. Connecting the cells in series increases the voltage, but does not affect the useful life of the cells.

Ⅱ.

1. e 2. h 3. f 4. g 5. a 6. b 7. c 8. d

Ⅲ.

interconnection; electric; active; generating; passive; inductors; voltage; deliver

Ⅳ.

1. 理想独立电源是一个能够提供特定电压或电流的有源元件，该电压或电流不依赖于电路中其他变量。
2. 电路元件吸收或释放的功率为元件两端的电压与流过该元件电流的乘积。
3. 电阻越大，通过导线输送一定电流所需的电压就越高。
4. 对于不同端电流而具有不同电阻值的电阻器被称为非线性电阻器。

阅读材料　　　　　　　　　　电气元件

1. 电阻器

电阻器是一种能阻碍电流流动的电子器件。在电阻器中流过的电流与加在电阻两端的电压成正比，与电阻的阻值成反比，这就是欧姆定律，用公式表示为 $I = U_R/R$。电阻器是线性器件，它的（伏安）特性曲线呈一条直线。

电阻器常用作限流器，限制流过器件的电流以防止器件因通过电流过大而烧坏。电阻器也可用作分压器，以降低其他电路的电压，如晶体管偏置电路，电阻器还可用作电路的负载。

2. 电容器

电能可以储存在电场里，储存电能的元件叫作电容器。

一个简单的电容器由被介质隔开的两块金属板组成。如果电容器连接到电池上，电子将从电池的负极流出堆积在与负极相接的极板上。同时与电源正极相接的极板上的电子将离开极板流入电池正极，这样两极板上就产生了与电池上相等的电势差。我们就说电容充上了电。

用一根导线连接电容器的两个极板，电容就会放电。电子从一个极板通过导线向另一个极板运动，去恢复电中性。

3. 电感器

当电流流过电感器时，电感器周围就有电磁场，电感器是以电磁场的形式暂时储存电磁能量的电子器件。电感器由不同尺寸的导线绕制而成，且有不同的匝数，这些都会影响线圈的直流电阻。

在美国无线电转播联盟出版的《业余无线电手册》中有满足所需的技术条件绕制线圈的详细资料。同时，有许多便宜的专用计算尺可以帮你设定所需要的参数，确定电感器的匝数、线圈长度、线圈直径等，这些都是达到预想结果所必需的。

Keys to Exercises of the Reading

Ⅰ.

1．F　2．F　3．T　4．F

Ⅱ.

1. Electrical energy can be stored in an electric field. The device capable of doing this is called a capacitor or a condenser.

2. A simple condenser consists of two metallic plates separated by a dielectric.

3. Inductors are wound with various sizes of wire and in varying numbers of turns which affect the DC resistance of the coil.

第 2 单元　交流电、直流电与电信号

1. 交流电

交流电流持续不断的转换方向，从一个方向转换成另一个方向（如图 2-1 和图 2-2 所示）。交流电压持续不断地在正负极性之间交替变化。这种变换方向的速率称为交流电的频率，测量单位是赫［兹］（Hz），它表示一秒内（交流电）正反向周期性变化的次数。

交流电源适用于给一些设施供电，比如灯和加热器，但是几乎所有的电子电路都需要一个稳定的直流电源。

2. 直流电

直流电流总是向相同方向流动，但它可能增大或者减小。直流电压总是正的（或总是负的），但它也可能增大或者减小（如图 2-3 所示）。电子电路通常需要一个具有定值的稳定的直流电源。干电池、蓄电池和稳压电源能提供对于电子电路来说理想的、稳定直流电。

在任何直流电源作用下，电灯、加热器和发动机都可以正常工作。

3. 电信号的特性

电信号是携带信息的电压或者电流，通常是电压。这个术语可以用于电路里任何电压或电流。图 2-4 的电压－时间图说明了电信号的各种特性。除了图上标示的特性外，还有频率，它表示（电信号）每秒钟的周期数。该图以一个正弦波为例，但这些特性适用于具有恒定形状的任何信号。

振幅是信号所能达到的最大电压。单位是伏［特］，用 V 表示。峰值电压是振幅的另一种说法。峰－峰值电压是峰值电压（振幅）的两倍。当对示波器轨迹进行读数时，通常测量峰－峰值电压。

时间周期是信号完成一个周期所需要的时间，测量单位是秒（s）。但是时间周期往往非常短，所以常用毫秒（ms）和微秒（μs）测量。1ms = 0.001s，1μs = 0.000001s。

频率是每秒的周期数，单位是赫［兹］（Hz）。但频率往往比较高，所以常用千赫（kHz）和兆赫（MHz）。kHz = 1000Hz，1MHz = 1000000Hz。频率 = 1/周期，周期 = 1/频率。

另一个常用的值是交流电的有效值，交流电压或电流有效值作用于电阻，与相同大小的直流电压或电流作用于相同大小的电阻等效。

Keys to Exercises of the Text

Ⅱ.
1. positive, negative
2. increase, decrease
3. maximum, volts, Peak voltage
4. signal, milliseconds, microseconds

Ⅲ.
1. D 2. B 3. A 4. B

Ⅳ.
一直沿着一个方向流动的电流通常叫作直流电。我们知道汽车和飞机的电子系统、电报、电话和无轨电车使用直流电。直流电也用来满足一些工业需求。然而，对工业和许多其他用途，城市中更多地使用另外一种电流，这种电流首先沿着一个方向流动，然后沿着另一个方向流动，它被命名为交流电。我们知道交流电就是使无线电成为可能的那种电流。

名人传记阅读材料（Ⅰ）　　　尼古拉·特斯拉

尼古拉·特斯拉于 1856 年出生于克罗地亚的斯米良利卡。他是一个西伯利亚东正教神父的儿子。特斯拉在奥地利工业学校学习工程学。他在布达佩斯做电气工程师，后来在 1884 年移民到美国后在爱迪生机器公司工作。他与 1943 年 1 月 7 日在纽约去世。

在他的一生中，特斯拉发明了荧光灯、特斯拉感应电机、特斯拉线圈、交流电供电系统，这个系统包括一个电机和变压器和三相电。

在最高法院于 1943 年推翻了古格列尔莫·马可尼的专利而判特斯拉早期专利有效后，特斯拉现在也被公认发明了现代无线电；当一个叫奥·斯·庞德的工程师谈论马克尼的无线电系统时对特斯拉这样说"好像马可尼在您的基础上起跳"，特斯拉答道，"马克尼是个不错的伙计。让他接着干吧。他使用了我的 17 个专利。"

特斯拉在 1891 年发明的特斯拉线圈，现在还使用在电视机和其他电子产品上。

尼古拉·特斯拉——神秘的发明

在发明了生产交流电的方法并申请专利 10 年后，尼古拉·特斯拉声称发明了一种不需要任何燃料的发电机。这个发明已经在公众面前丢失了。特斯拉声称他的这个发明可以利用宇宙射线来驱动动力装置。

特斯拉被授予了超过一百件专利，还有很多数不清的未申请专利的发明。

尼古拉·特斯拉和乔治·威斯汀豪斯

1885 年，威斯汀豪斯电气公司的老板乔治·威斯汀豪斯购买了特斯拉的发电机、变压器和电动机的专利权。威斯汀豪斯使用了特斯拉交流电系统来给 1893 年在芝加哥召开的哥伦布纪念博览会提供照明。

尼古拉·特斯拉和托马斯·爱迪生

尼古拉·特斯拉是托马斯·爱迪生在 19 世纪末的竞争对手。实际上他在 19 世纪 90 年代比托马斯·爱迪生更出名。他的多项电发明为他在世界上赢得了很高的知名度和大量财富。在他的人生顶峰时期他是很多成功人士的密友，这些人包括诗人、科学家、工业家和银行家。然而特斯拉去世时贫困潦倒，丢失了财富和在科学界的声誉。在他从声名远扬到变得默默无闻的下降过程中，特斯拉留下了这些名副其实的发明遗产给后人，还有现在人们还在为之着迷的一些预言。

Keys to Exercises of the Reading

Ⅰ.

1. T 2. F 3. T 4. T

Ⅱ.

1. Tesla claimed that he worked from 3 a.m. to 11 p.m., no Sundays or holidays excepted.

2. After 1890 Tesla experimented to transmit power generated with his Tesla coil by inductive and capacitive coupling with high AC voltages.

3. Trying to come up with a better way to generate alternating current Tesla developed a steam powered reciprocating electricity generator which he patented in 1893 and introduced at the Worlds Colombian Souvenir Exposition that year.

第 3 单元　电　子　仪　器

1. 万用表

万用表是一种通用仪表，可用来测量直流和交流电压、电流、电阻，有的还能测量分贝（放大倍数的情况）。有两种万用表：一种是用指针在标准刻度上的移动来指示测量值的模拟万用表（见图 3-1a），另一种是用电子数字显示器显示测量值的数字万用表（见图 3-2b）。这两种万用表都有一个正极（＋）插孔和一个公共端（－）插孔用来插入测试笔，一个功能选择开关用来选择（测量对象）：直流电压、交流电压、直流电流、交流电流或电阻（欧姆值），一个范围选择开关用来（选择范围）以测出精确读数。万用表还可能有其他插孔用来测量高电压（1～5kV）和大电流（高达 10A）。对一些特殊的万用表而言，还会有一些其他功能上的变化。

除了功能选择开关和范围选择开关（有时它们合并为一个开关），模拟万用表可能还有一个极性开关，可以很方便地交换测试笔的极性。指针常常有一个旋钮来机械调零。另外当测量电阻时，一个零点调节控制（钮）用来对电池电压不足时做补偿调节（即保证电阻为 0 时指针指向零值）。模拟万用表可以测量正电压和负电压，只要简单地对调一下两个测试笔或拨一下极性开关。数字万用表通常会自动在显示器上指示出极性。

为了确保读数正确，万用表必须与电路正确连接。一个电压表（万用表测量电压时）应与被测电路或元器件并联。当测量电流时，电路必须断开，插入万用表表笔，使万用表与被测电路或元器件相串联。当测量电路中局部电路（或元器件）的（等效）电阻时，必须关掉电路中的电源，使万用表与该局部电路（或元器件）并联。

2. 示波器

示波器（见图3-2）是一个图像显示设备，它显示一个电子信号的图像。当信号输入到示波器中时，就产生一个电子束，该电子束被聚焦、加速并适当偏离，在阴极射线管的显示屏上显示电压的波形。

示波器通常显示电压信号如何随时间变化的图像：其纵轴 Y 表示电压，横轴 X 表示时间。在示波器屏幕上的电压波形的幅度可以通过数出电压波峰与波谷之间的纵向距离来确定（见图3-3），将这个值乘上 V/cm 控制钮的设定值就得到电压的幅度值。比如说，假若电压的峰-峰值幅度为5cm，控制钮设在1V/cm处，那么，峰-峰值电压为5V。

用示波器的水平标尺可以测量时间值。时间测量包括测量信号的周期、脉冲宽度和频率。频率是周期的倒数，所以一旦知道了周期，频率就是用1除以周期即可得到。

一个波形的频率可以通过在水平方向数出该波形一个周期的长度来确定，将这个值乘上 t/cm 控制钮的设定值就得到它的一个周期所需的时间。例如，假定一个波形长4cm，控制钮设在1ms/cm，那么周期是4ms，则频率可以用下面的公式算出：

$$f = \frac{1}{P} = \frac{1}{4\text{ms}} = 250\text{Hz}$$

假定控制钮设在100μs/cm，则周期是400μs，频率为2.5kHz。

双踪示波器具有同时显示输入信号和输出信号的优点，可以显示输出信号是否有失真和表明输入/输出信号的相位关系，即两路信号的波形重叠在一起可以较好地显示出输入与输出信号相位的漂移。

Keys to Exercises of the Text.

Ⅱ.
1. 通用仪表   2. 交换测试笔
3. 机械调节   4. 电压振幅
5. 双踪示波器   6. 信号发生器
7. 模拟万用表   8. 相位关系
9. 显示电压波形   10. 正电压

Ⅲ.
1. The analog multimeter may have a polarity switch to facilitate reversing the test leads.
2. There are some variations to the functions used for specific meters.
3. The oscilloscope displays/ draws a graph of an electrical signal.
4. A dual-trace oscilloscope is advantageous to show the input signal and output signal of one circuit in the same time.
5. The two traces may be placed over each other (superimposed) to indicate better the phase shift between two signals.

阅读材料　　　　　　　如何使用测试仪表？

该仪表是设计用来测量直流电压、交流电压、电阻、导电性能和测试二极管的。该仪表

具有 $3\frac{1}{2}$ 位数字的液晶显示，因此它也被称作数字多用表。

该仪表的使用方法示例如下：

Ⅰ．直流电压的测量（见图 3-4）

1）功能开关置于 DCV。

2）测试表笔接在被测电路上。

3）读显示值。

注：

1）表笔极性反接时，会出现"－"（负号）。

2）在测量包含有尖峰脉冲的电压时（如电视中的水平输出信号）使用正接的测试表笔。

Ⅱ．交流电压的测量（见图 3-5）

1）功能开关置于 ACV。

2）测试表笔接在被测电路上。

3）读显示值。

注：不用考虑测试表笔的极性。

Ⅲ．电阻测量（见图 3-6）

1）功能开关置于 Ω 档。

2）测试表笔接在被测电路上。

3）读显示值。

注：接表笔之前要确认已断开被测电路的电源。

Ⅳ．导电性能测试（见图 3-7）

1）功能开关置于符号为"·))"上。

2）表笔接在被测试电路上。

3）有蜂鸣音且显示标识"·))"时，表示导通良好。

Ⅴ．二极管测试（见图 3-8）

1）功能开关置于符号为"·))"上。

2）对正接的二极管，当黑表笔接其阴极，红表笔接阳极时，显示为正向电阻值；当测试表笔接反时，显示数字为"1."。

3）当测试表笔开路时，显示读数为"1."。

Keys to Exercises of the Reading.

Ⅰ．

1. This instrument is designed to use for measuring voltage, measuring resistance, conductivity test and diode test.

2. The following is illustrating its operation.

3. Minus sign "－" is displayed when the polarity of the test leads is reversed.

4. Connect the test leads to the circuit to be measured.

第 4 单元　集成电路

集成电路是几种相互连接的电路元器件，如晶体管、二极管、电容器和电阻器等的组合。它是用半导体材料制成的小型电子器件。第一块集成电路是在 20 世纪 50 年代由德州仪器的 Jack. Kilby 和创意半导体公司的 Robert·Noyce 合作开发的。

电学上把组成集成电路的彼此相连接的元器件称为集成元器件。如果集成电路只包含一种类型的元器件，便称为元器件组件。

在各种设备包括微处理器、音频和视频设备以及汽车中都要用到集成电路（见图4-1）。集成电路通常根据其包含的晶体管和其他电路元件的数量来归类。

· SSI（小规模集成电路）：每个芯片中有 100 个以下的电子元器件。
· MSI（中规模集成电路）：每个芯片中有 100 到 3000 个的电子元器件。
· LSI（大规模集成电路）：每个芯片中有 3000 到 100000 个的电子元器件。
· VLSI（超大规模集成电路）：每个芯片中有 100000 到 1000000 个的电子元器件。
· ULSI（甚大规模集成电路）：每个芯片中有 100 万个以上的电子元器件。

随着在一个芯片上集成大量晶体管的能力（即集成电路的集成量）的提高，对专用集成电路的需求已更加普遍。至少对大批量的应用来说更需要专用的集成电路。硅芯片技术的发展使得集成电路设计者可以在一个芯片上集成几百万个以上的晶体管，甚至现在可以在单一芯片上集成一个中等复杂的系统。

集成电路的发明是电子工业的一次重要革命。凭借这一技术，完全可以实现尺寸缩小和重量减轻，更重要的是能够实现高可靠性、良好的工作性能、低成本和低功耗。集成电路已广泛应用在电子工业中。

Keys to Exercises of the Text.

Ⅱ.

1. T 2. T 3. F 4. T 5. F 6. F

Ⅲ.

1. An integrated circuit (IC) is a combination of a few interconnected circuit elements on a piece of semiconductor chip.

2. The electrically interconnected components that make up an IC are called integrated elements.

3. ICs are widely used in the electronic industry.

4. Integrated circuits are often classified by the number of transistors and other electronic components

Ⅳ.

集成电路看起来只不过是一个微小的金属片，或许一边只有 1/2 厘米长，比一张纸厚不了多少。芯片这么小，如果它掉到地板上，就很容易跟灰尘一起扫走。尽管它很小，芯片的生产过程却应用了当代最先进电子技术。到今天的发展水平，芯片可能包含一万多甚至数百万独立的电子元器件（这些电子元器件有许多不同的功能，诸如二极管，晶体管，电容器和电阻器等）。

第 5 单元 运算放大器

运算放大器是所有线性电路中最重要的基本器件，其应用十分广泛，包括模拟信号的运算、放大、滤波、产生或进行各种线性或非线性的处理等。

运算放大器这一术语最早适用于下限频率可低到 0Hz 的高增益放大器。它可用于模拟计算机中执行某些数学运算。高增益的放大器现在已广泛应用于各个方面，早就远远超出了数值运算的范围，但通常仍称为运算放大器或 op-amp。早期的运算放大器采用分立元件，但现在使用集成电路就更加方便了。电路设计者通常对集成电路的内部元件不感兴趣,只关心作为一个整体的单元性能。所以图 5-1 所示的符号用来表示运算放大器。由图可以看出,运算放大器有两个输入端,一个输出端以及正、负电源端子。

假如反向输入端正电压略微增加一点,那么输出负电压将变得更强,这就是之所以称这个输入端为"反向输入端"的原因。然而,假如同相输入端电压正向增大,输出端正电压也将变得更强。

741 型器件是众所周知的通用运算放大器之一,也是所有的线性集成电路中最廉价的一种。这种器件实际上具有很多不同的封装形式。图 5-2 所示的 741 型器件是最常见的,这种封装形式为 8 引脚双列直插式。

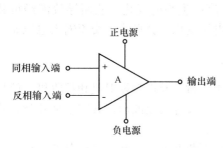

图 5-1 运算放大器

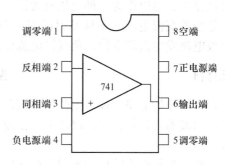

图 5-2 741 运算放大器

741 反相放大器的基本电路如图 5-3 所示。输入信号经 R_1 加在 741 器件的反相输入端,因此输出信号与输入信号的相位相反。运算放大器被设计成具有很高的输入阻抗,所以 741 的净输入电流很小。因为反相输入端虚短,流过 R_1 的电流等于 U_i 除以 R_3,同样流过 R_3 的电流等于 $-U_o/R_3$,其中 U_o 是输出电压。这两个电流几乎相等,因此电路的电压增益等于电阻值:

$$G = -\frac{R_3}{R_1}。$$

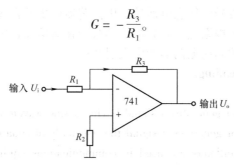

图 5-3 741 反相放大器

Keys to Exercises of the Text.

Ⅱ.

1. filtering waveforms   2. operational amplifier
3. the input impedance   4. the positive and negative supply lines
5. the inverting input is virtually at ground potential
6. the performance of the unit as a whole
7. the 8 pin dual – in – line integrated block

8. performing linear and non – linear signals handling

Ⅲ.
1. 程序的作用是将用户的意愿传递给计算机。
2. 最好的方法是以非常快的浏览器去访问万维网。
3. 开（要）启一个 NPN 双极晶体管，基极一定比发射极电压更高（才行）。
4. 多媒体系统来支持多种媒体类型是必要的。

Ⅳ.
放大器在诸如收音机、示波器和录放机等许多电子设备中是必要的。它常用于放大一个微小的交流电压。共发射极模式的结型晶体管，如果负载（大小合适的电阻器）连接到集电极回路上，可以作为一种电压放大器。

把小的交流电压 u_i 加到基射回路，引起基极电流产生小的变化，从而导致流经负载的集电极电流发生大的变化，负载将电流变化转变成电压变化，就形成了交流输出电压 u_o。

阅读材料　　　　　　　　　振　荡　器

产生信号的电路被称为振荡器。振荡器广泛应用于射频及视频发射机和接收机、信号发生器、示波器和计算机中，以产生正弦波、方波、锯齿波等交流波形，其频率从几赫到几百万赫。

因此，根据产生波形的不同将振荡器分为两类：正弦波振荡器和张弛振荡器。正弦波振荡器产生正弦波形的信号，而张弛振荡器通常产生方波信号。正弦波振荡器广泛用于无线电通信中产生载波，而且在许多测试仪器中使用该载波系统。张弛振荡器在电视及雷达系统、数字系统和测试仪器中用作脉冲发生器。

振荡器通常包括放大器和某种类型的反馈系统（反馈是指将输出信号送回至放大器的输入端）。选频元器件可以是电感－电容调谐电路或振荡晶体。晶体控制振荡器的精确度和稳定性非常高。振荡器多用于产生音频及射频信号。例如，用于现代按键电话中的简单音频振荡器能传送数据到需要拨号的任意电话站。由振荡器产生的音频声音还用于闹钟、无线电通信、电子琴、计算机和警报系统。高频振荡器被用在通信设备中实现调谐和检波功能。广播电台和电视台应用精密的高频振荡器产生发射频率。

Keys to Exercises of the Reading.

Ⅱ.
1. A sinusoidal oscillator generates a signal having a sine waveform.
2. A relaxation oscillator generates a signal that is usually of square waveform.
3. High – frequency oscillators are used in communications equipment to provide tuning and signal – detection functions.
4. The frequency – determining elements may be a tuned inductance – capacitance circuit or a vibrating crystal.

第6单元　强电工程和弱电工程

重型机器制造业都需要有一个动力源来进行工作。它还要有一个动力分配网，以便把能量输送到该系统的各个部分。但系统的功率流具有双重目的。

1）给系统提供动力，使其能够完成特定任务。
2）为了在系统的各处输送数据，能量在这里作为传送数据的载体。

许多年前，我们在电气工程中常把电力工程（强电工程）和弱电工程区别开来。前者主要同电力的生产、输送和利用有关；而后者则包含了电子学、电话、无线电等诸如此类的一些方面。这样划分的理由部分是由于弱电工程主要和通信和控制工程相关，而这两方面仅需要少量的电量来支撑。然而随着时代的变化我们认识到，通信也是要依靠能量来进行的，并且距离越远、需要传送的信息量越大，则对电力的需求也随之增加。

举一个具体例子来说明，如果一个教师给一个仅有一二十名学生的班级授课，他只要用正常的音量就足够了；但是如果这个班级人数增加到大约五十或一百名学生时，他就需要提高音量，这样当授课结束时，他或许会感觉到疲惫。如果听众多达几百人或者几千人，那么讲话者为了使听众能够听到他的声音，并顺利接收他传出的信息，就需要大功率的扩音机来提高他的音量。如此一来，距离越远，所需要传送的信息量越大，则对电力的需要量也就越大。

如今一台常见的无线电发射机的功率已达 100kW，因此它的播送和影响范围很大。电视和其他大众传媒工具也是如此。

电气工程自出现以来，已经有了很大的发展。现在我们通信系统中的发报机，为了完成指定任务，功率已能够高达几千千瓦。

雷达又是一个例子。在探讨雷达问题时，不得不说雷达的覆盖空间很大，而且要求在很短的时间内提供大量准确的数据。也正是由于这些任务的浩繁且要求严密，雷达系统所需要的功率可高达很多兆瓦。

如此一来，强电工程和弱电工程在这一点上的区别已经不明显了。但是我们认为它们在概念上还是有所不同的，电力工程主要关注地是在远距离空间内输送能量，而通信系统的主要目的则是传递、提取和处理数据，尽管在此过程中也可能消耗大量的电力。

Keys to Exercises of the Text

Ⅰ.

1. F 2. F 3. T 4. T 5. F

Ⅱ.

1. j 2. c 3. b 4. g 5. d 6. e 7. f 8. h 9. i 10. a

Ⅲ.

1. The more air there is inside the tyre, the greater the pressure there is in it.

2. Nowadays, a typical radio transmitter has a power of 100 kilowatts so that it can broadcast information over a large area of influence.

3. It is because of the multitude and the exactitude of this project that the cooperation between the two countries should be strengthened.

4. We have progressed a long way from the early days of electrical engineering.

Ⅳ.

电气工程自出现以来，被广泛应用于我们的现实生活中。例如，一台常见的无线电发射机可以在很广的影响范围内播送信息。雷达系统也是一种常见的应用，当然还有电视和其他大众传媒工具。

第7单元 电力网是如何工作的

电力网有点儿像你呼吸的空气，除非没有了，否则你不会想到它。电一直在"那里"满足着你的需要，只有当停电时，你走入黑暗的房间里本能地想去打开无用的开关时，才会发现电在你的日常生活中是多么重要。你用电来取暖、制冷、煮饭、冷藏食物、照明、计算和各种娱乐等。没有电,生活就会有许多不方便的地方。

电从发电厂传到你家(如图7-1所示)是通过一个奇妙的系统,称为电力(分布)网。电力网很普遍,下面我们来看一下把电能从发电站送到你家的各种设备。

1. 电厂

电从电厂开始,几乎所有的电厂都包括旋转式发电机。带动发电机旋转的设备在水力发电站可能是一个水轮(机),也可能是一个大的柴油机或一个汽轮机。但大多数情况下是蒸汽轮机带动发电机旋转。蒸汽可能通过燃烧煤、石油、天然气,或者通过核反应堆产生。无论是何种方式拖动发电机,各种型号的(商用)电力发电机都产生三相交流电。为了理解三相交流电,先看单相交流电。

2. 交流电

单相交流电就是在家里所用的电源。通常提到的民用电就是单相220V交流电源。如果用示波器看家里墙上电源插座的输出电压,你会在示波器的显示屏上看到电源的波形像一个正弦波,这个正弦波在-311V和311V之间振荡(峰值确实是311V,有效值电压是220V)。正弦波振荡的速率是每秒50次。这种振荡电源通常称为交流电。

3. 电厂产生交流电

电厂同时产生三相交流电,三个交流电的相位互差120°。每个电厂出来四根电力线,三根相线加上一根中线或三相的公共接地线。三相交流电相对于地的电压波形如图7-2所示。

4. 输电变电站

从发电机出来的三相电进入电厂的输电变电站。变电站用大变压器把发电机的电压(大约几千伏)升压为极其高的电压,为在电力网上的长距离输出做准备。

5. 配电网

在家庭或商业中使用的电,由输电网的电逐级降压进入配电网,可能分成几个阶段。把电从输电网转换到配电网的地方是变配电站。

6. 配电总线

从变压器出来的电流入配电总线:在这种情况中,配电总线的电力按两种不同电压分成两组配电线,与总线相连的小变压器把电压降低到标准的线电压(7200V),主变压器出来的较高电压的电力线则朝另一个方向传送。电力线按三线一组分成两组,朝着不同的方向离开这一变电站。

在电线杆顶部的三根线是三根相线,略低一点的第四根线是地线,有时还有一些附加的线如电话线、有线电视线等安装在同一根电线杆上。

7. 支线及进屋

民用的只是三相电中的一相,所以通常你会看到沿着主干道的三相线和通向小街小巷的一相或二相支线。变压器把7200V电压降为220V家用标准电压,电力线通过一个电度表后把220V的交流电源送到你家。

Keys to Exercises of the Text

Ⅱ.

1. cooling, computation
2. natural gas, nuclear
3. generator, transmission substation

4. transformer drum

Ⅲ.
minimizing; operation; security; maintaining; make up; damaged; transmission; limit

Ⅳ.
由于电能不能以简单、经济的方式大量存储,电能的产生和消费必须同时完成。在电力系统任何层次上的故障或误操作都可能导致对用户供电的中断。因此,保持电网的连续正常运行对给客户提供可靠的电能供应是极其重要的。

阅读材料　　　　　　　　　　　　　　变　压　器

变压器大小不同。有的变压器像房子一样大。而电子变压器可能小如糖块。所有的变压器至少有一个线圈;大多数变压器具有两个以上的线圈。

通常使用变压器的目的是改变电压等级,但有时也用于隔离电源和负载。

1. 变压器的类型

标准电力变压器有两只线圈。这些线圈标记为一次绕组(一次侧)和二次绕组(二次侧)。一次绕组连接电源,二次绕组连接负载。在一次和二次绕组间没有电互连。二次绕组通过感应获得电压。

大概只有在发电厂才能见到升压变压器。典型的发电机出口电压为13800V,升压至345000V用来输送。下一站是变电所,把电压降到配电电压等级,约15000V。大型的变电所变压器具有冷却叶片,以防止变压器过热。有的变压器安装的位置靠近用户。

2. 变压器的结构

变压器的线圈之间是绝缘的,但存在磁联系。两只线圈绕在同一个铁心上。一次绕组电流使铁心磁化,并在铁心中产生磁场。铁心磁场影响一次和二次绕组的电流。

铁心主要有两种型式:

1)心型。铁心在线圈内部。
2)壳型。铁心在线圈外部。

小型电力变压器通常为心型。特大型变压器为壳型,但是在运行时二者并无区别。

线圈用铜线绕成。电阻应尽量小,从而保持低损耗。

3. 理想变压器

变压器效率很高,损耗通常低于3%。这使我们假定它在计算时是无损耗的。

无损耗意味着线圈没有电阻,并且铁心中没有能量损耗。进一步可以假定没有漏磁,即所有的磁力线穿过线圈的每一匝。

4. 励磁电流

欲知损耗的大小,可以看一下励磁电流。假定二次绕组空载,在一次绕组施加额定电压后会产生一较小电流。典型情况是该励磁电流小于额定电流的3%。

励磁电流由两部分组成。一部分与电压同相,该电流供给铁心中的能量损耗。铁心损耗的原因是涡流电流和磁滞。

涡流电流在铁心中旋转,由感应现象产生,因为铁心毕竟是处于变化磁场中的导体。

磁滞损耗是由用来排列铁心中的磁畴所需要的能量产生的。这种排列调整是连续不断的,开始是一个方向,然后是另一个方向。

另一部分励磁电流磁化铁心。磁化电流供给"往复能量",往复能量就是储存在磁场中的能量,并在每个周期内返回电源两次。磁化电流与所加电压正交(相位差为90°)。

Keys to Exercises of the Reading

Ⅱ.

1. The coils of a transformer are electrically insulated from each other. There is a magnetic link, however.

2. The core field then affects current in both primary and secondary.

3. To get an idea of just how small the losses are, we can take a look at the excitation current.

第 8 单元　继电器的基本类型

可以有许多种方法对继电器进行分类，如按功能、构造和用途等进行分类。从构造上来讲，继电器分为两类：电磁继电器和固态继电器。电磁继电器依靠电磁力作用于可动部件，从而实现开关机能，达到闭合或断开一系列触点的目的。固态继电器不是通过物理运动，而是通过将一系列固态元件从非导通转换为导通（反之亦然）来实现开关切换功能。电磁继电器历史较长并广泛使用，固态继电器更通用，并且具有潜在的稳定性，速度很快。

从结构上，继电器可分类如下：

1) 可动衔铁式继电器。可动衔铁式继电器如图 8-1a 所示。线圈 C 绕在磁性材料制成的磁心 M 上。在可动衔铁 A 上带有一个可动触点 B。在线圈中无电流通过时，触点 B 将 1-1 接点相连接。衔铁在重力或弹簧的作用下保持在图示位置。如果流过线圈中的电流缓慢增加，当到达一定阈值时，衔铁 A 会突然旋向左边，断开 1-1 接点并闭合 2-2 接点。能够起动衔铁运动的最小电流被称为始动电流。如果此时线圈中流过的电流缓慢减小，将会发现存在

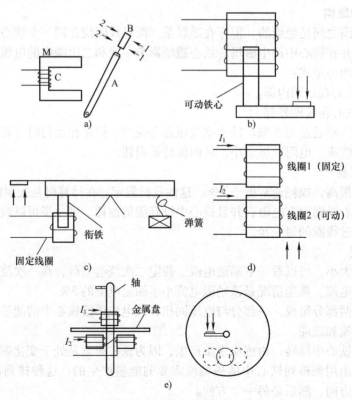

图 8-1　电磁继电器的类型
a) 可动衔铁继电器　b) 螺线管继电器　c) 平衡杆继电器　d) 线圈继电器　e) 感应圆盘式继电器

一个电流值,使得衔铁返回原来的位置,这称为释放电流。

2)螺线管继电器。如图 8-1b 所示,固定线圈环绕着可动的铁心,该铁心通过重力或弹簧的作用从线圈中的平衡位置移开。当线圈中的电流足够大时,铁心就被吸入线圈中,从而通过位移合上触点。同样对于这种继电器始动电流大于其释放电流。在任何选定的位置上,可动铁心所受到的力都与电流的平方成正比(或者与线圈两端的电压的平方成正比)。

3)平衡杆继电器如图 8-1c 所示,衔铁固定在平衡杠上,当线圈中流有电流时,衔铁就被拉入固定的线圈中。这种继电器可以采用各种触点和弹簧结构。

4)利用线圈间相互作用构成的继电器。两个线圈,一个是固定的,一个是可移动的,如图 8-1d 所示那样布置,使它们的磁场相互作用。如果线圈中流过的电流方向不变,两个线圈间的作用力正比于两个电流的代数积,并且会将两个线圈吸在一起或因相互排斥而分开。如果线圈中流过的是交流电流,相互间的作用力将正比于这两个电流的大小以及它们之间相位差的余弦的乘积,即

$$F = K_M I_1 I_2 \cos\Delta$$

5)感应圆盘式继电器。感应圆盘式继电器的工作原理与感应电动机类似。如图 8-1e 所示,时变电流流经空间位置变化的线圈导致移动磁场,该磁场穿过可动圆盘,在圆盘中激发出感生电流,(这个感生电流在磁场中)会使圆盘受到力的作用,带动圆盘旋转。在这类继电器中,圆盘上方的线圈中流有一个电流,在圆盘下方的线圈中流有另一个独立电流。如果这两个电流相互之间的相位差为 Δ,则对圆盘产生的力矩为

$$T = K_D I_1 I_2 \sin\Delta$$

如果这两套线圈是相互串联的,因此流有相同的电流,则无力矩产生。但是通常采用在一组线圈两端并联一个电阻或电容的方法使该线圈中的电流的相位不同于另一组线圈中的电流,从而产生力矩。当然也可以采用其他方法来移相。感应继电器的线圈通常提供抽头,用来选择不同的始动电流。这样,即使在多年使用中负载的布局发生了改变,仍能将保护调整得与之相适应。

Keys to Exercises of the Text

II.

1. F 2. F 3. F 4. T 5. T

III.

1. electromagnetic relay
2. solid state relay
3. moving – armature relay
4. release current
5. pick – up current
6. solenoid relay
7. balanced beam relay
8. mutually interacting coil
9. induction disc relay
10. movable disc

IV.

1. All bodies of matter conduct heat to some degree. But many solids conduct heat better than liquids, and liquids better than gases.

2. Planets revolve about the sun and rotate on their axis.

3. Electric current is directly proportional to voltage and inversely proportional to resistance.

4. These bad conditions have given rise to a lot of crime.

第9单元 电力系统监控

现代供电系统是许多部分的组合，每个部分都影响其他部分的运转。为使系统作为整体正常运作，有必要监控系统中许多不同位置的情况，才能确保最佳运转。

用户主要关心的是把频率和供电电压限制在某个相当有限的范围内。因为在任何地方系统的频率都相同，所以可在任一方便的地方安装单个的频率测试仪来监控。相反，系统的电压可能在不同的地方差别很大。因此，必须在系统某些关键位置对电压进行不间断地观察，才能提供令人满意的服务。

将合适的负荷计划分配给系统中每台发电机，才能使系统达到高效运转。新的设备虽然单机效率较高，但它们在系统中被安装的位置可能会由于它们的加载而造成系统巨大的损耗。应该将负荷在发电机之间分配好后进行运作，才能使燃料总消耗降至最低。为在意外情况下可靠供电，最好使运行中的所有机器的千瓦定额比负荷加上损耗的总数高一些。

要想把用户的用电量记入账目，就必须安装测试设备。在不同的供电系统间存在着许多互连。必须在连接点配备测量仪器才能将从一个系统转换到另一个系统的能量记入账目。必须对能量转换不断监测，才能保证交换的电力在协议规定的限度之内。

要避免由于超载造成的损害，必须对大件装置的状态不断测量。测试设备在未来扩充供电系统的建设中起指导作用。

有时在紧急情况下，一位系统操作人员注意到他的系统负荷超过了现有发电和输电设备的能力。于是他就面临着限制用电的问题，更恰当地说，是保持负荷的问题。于是就有必要在最合适中断供电服务的地方减少一定的负荷。在此情况下，他依赖于许多测试仪表提供的有关系统运行状况的信息。

当要求采取操作避免超过设计限定运行给设备带来损害时，测量仪会发出警报声，作为预警。万一出现最糟的情况，如电力系统出现故障，故障设备将会自动关闭，停止运行。持续监控电流、电压和其他量的测量仪器，必须能够辨别出产生故障的设备，使断路器开始动作将有故障的设备排除在供电服务之外，而让运行系统中其他所有设备继续运行。

电力系统和用户拥有的许多不同的电力装置都是为在某些特定范围内运行而设计的。不可以在设计限定范围之外运行，因为它会导致低效运行、过度磨损或者（在极端的情况下）设备损坏。认真观察设备运行状态才会知道应该如何采取正确措施。

所有电力装置都不可以过流，因为这会导致温度过高、低效运行并减少使用寿命。居民电路若电流过载，会导致因熔丝烧断而产生断路，或使断路器动作。电动机若电流过载，会损坏绝缘，使绝缘失效。

Keys to Exercises of the Text

Ⅱ.
1. F 2. T 3. F 4. F 5. T

Ⅲ.
1. frequency meter 2. electrical power transmission equipment
3. power system fault 4. circuit breaker
5. service interruption 6. system load
7. load shedding 8. electrical equipment

Ⅳ.
1. The car speed must not exceed a maximum of 55 miles an hour.

2. A business must monitor changes and needs in society in order to behave in a social responsible way.

3. To meet these challenges, most importantly, we should rely on development.

4. Attention should be focused on what is needed to obtain the required answer.

阅读材料   **配电自动化可增加可靠性**

长岛照明公司凭借750条附加配电线路，主要是高架线路，向100多万用电单位提供供电服务。这些供电线路易受到与暴风雨有关问题的侵扰，包括雷暴雨、闪电、树挂线、暴风雪及飓风。虽然用户常受到长时间的停电干扰，但与纽约其他电力公司相比，一般来讲，这家公司恢复供电时间最快。

在用户断电分析中，测出78%断电的原因是由三相线路上的故障造成的。平均每次主干线路故障将导致2000个用户停电。为了减少这些故障造成断电的次数，公司制定出完善可靠的方案，包括树木剪修、安装避雷器、避雷线以及替换老化的、失去绝缘性的绝缘器。这些方案的出台减少了停电次数，但还不能达到连续供电的理想水平。为进一步提高可靠性，长岛照明公司安装了先进的配电自动控制系统，该系统是由长岛照明公司与加拿大加尔伯达省卡加立市的哈里森配电自动化生产厂家共同开发的。该套系统是根据电压、电流、断路情况、监控开关状态的实时参数来隔离故障，并恢复主干线路中未受破坏部分的电压。自该系统在1993年被引用以来，24万多用户免去了困扰已久的断电干扰。

早在20世纪70年代，长岛照明公司就在电杆上安装了400多个无线电控制开关，这种开关包括中性介质的电压真空开关和电流传感器，用以减少停电持续时间，但对停电次数的减少作用不大。自动区域性断电需要一个带有本地智能性的遥控数据终端，用于测出永久负载侧故障，并在变电站出现故障之前，打开电动机控制开关。为达到这样的操作程序，选用了哈里森DART遥控数据终端。该系统能够在没有发送器的情况下，直接向各传感器输入交流信号。此外DART公司还拥有故障探测系统，此项技术已获专利，该系统是由美国伊利诺伊州芝加哥施恩禧公司生产和测试的。它能测出3个层级的过流，包括无电压损耗的过流、一次性断路跳闸关闭的过流和断路器闭锁的过流。

在故障检测系统的基础上，哈里森公司又开发了一种称为"SMART"开关或自动断路器的先进RTU系统。这种先进设备的使用不需要通信网络或继电器配合。把自动断路器接在400个现有开关上，该项任务是在1996年4月正式完成的。另外长岛照明公司还安装有350个具有自动切断功能的施恩禧SCADA型开关。这750个故障自动切断开关，使得长岛照明公司减少了25%的用户再受主三相故障造成的停电影响。共有24万用户在长达18个月的时间内免去了持续断电的干扰。

第二阶段的项目包括装配基于PC的哈里森SCADA系统和设置故障自动切断开关通信装置，它们可使操作者对开关实行远距离监视和控制。哈里森PC SCADA系统以三相电为基础，提供电压、电流、功率因数、负荷曲线以及其他实时数据，以优化这种配电系统的使用。在对一些通信手段进行了一番评估后，长岛照明公司选用了微波数据系统扩频通信网络。这种微波数据系统还混入了自由频段的扩频跳频技术。借助于使用微波数据系统，长岛照明公司安装了750个无线电台，在9个月的时间内通信覆盖面达90%。

第三个阶段的任务要求生产一种自动恢复到无故障供电线路的先进算法。该算法设计运行在哈里森D-200上，它（哈里森D-200）作为前端处理器用于与RTUS保持通信联系。

哈里森D-200从自动切断开关处获得电压损失和故障位置的信息。另外，哈里森D-

200 还获取变电站断路器情况和受影响线路的负荷以及邻近线路信息。之后，依照实时电压、负载、断路器情况、开关位置和安全闭锁情况，该算法计算和报告出恢复供电的正确方案。这种自动恢复算法能同时恢复 12 条供电线路，还可支持多达 7 个装在开环线路上的开关。此外，操作者能为每一条环路设置自动化等级。这些等级包括如下内容：

1) 手动模式：PC SCADA 系统不发出恢复供电指令，操作人员可完全监控和指示开关。

2) 操作者指令模式：哈里森 D-200 确定故障范围，然后根据实时信息发送给 PC SCADA 一个模拟图像以及恢复供电所需的操作步骤。如果操作人员同意算法推荐（操作），操作者就发出指令，使 PC SCADA 执行必要的操作步骤。任何步骤失败，都会中止恢复供电，从而使线路转入手动模式。

3) 全自动模式：根据算法确定的故障范围，自动恢复无故障线路的供电。一切分支步骤都是按实时信息数据来确定的，执行程序阻止工作区域通电。不需要人工介入。

1996 年 4 月 30 日，长岛照明公司获得这一先进的供电恢复算法的专利。但是许多公司仍采用管理开关来隔离故障，而且大多数采用操作员远程控制来恢复供电。这一程序耗时而且效率不高，特别是在暴风雪天气，多处出现线路故障的时候更是如此。

Keys to Exercises of the Reading

Ⅱ.
1. 避雷装置　　2. 持续断电　　3. 永久负载侧故障　　4. 多故障线路
5. 完全监控　　6. 环路　　　　7. 主动分断开关　　　8. 远程终端设备
9. 实时参数　　10. 三相电为基础

第 2 章　楼宇智能化技术

第 10 单元　智 能 建 筑

智能建筑将建筑管理与 IT 系统成功融合到优化系统性能和简易设备操作上，通过对建筑的结构、系统、服务和管理等要素进行最优化的分析后，设计提供一个安全、舒适、节能、高效、方便的工作环境；是传统建筑学与现代 4C 技术（计算机技术、控制技术、通信技术、图形显示技术）相结合的完美产物，已成为 21 世纪建筑业的发展主流。

智能建筑由楼宇自动化、通信网络自动化、办公自动化、综合布线系统、系统集成、安全防范自动化系统、消防自动化系统等组成，达到安全、舒适、节能、高效、环保的目的。

建筑设备自动化系统是以计算机为基础，监视建筑设备控制子系统的协调、组织和最优化，比如安全防范、火警/生命安全、电梯等。它主要包括暖通空调监控系统、建筑给水排水监控系统、建筑供配电控制系统、照明监控系统、交通监控系统、消防与安全防范系统、建筑设备自动化系统集成。应用如下：

1) 设备调度（根据需要开、关设备）
2) 启停优化（智能启动冷、热设备，确保在居住期建筑物处在所需要的温度范围）
3) 操作员调节（访问操作员设定点，把系统调整到要改变的状态）
4) 监视（温度记录、能源使用、设备开启时间等）
5) 报警（通知操作员设备故障、温度/压力超标或者需要维护）

通信网络系统包括通信网络系统和信息网络系统。通信网络系统是建筑物内的语音、数据、图像传输的基础设施，通过通信网络可实现与外部通信网路（如公用电话网、综合业务数字网、计算机互联网、数据通信网及卫星通信网等）相连，确保信息通畅和实现信息共享。信息网络系统是应用计算机技术、通信技术、多媒体技术、信息安全技术和行为科学等先进技术和设备构成的信息网络平台。

办公自动化系统由多功能电话机、高性能传真机、各类终端、PC、文字处理机、主计算机、声像存储装置等组成。办公自动化系统通过数字化创建、收集、储存、处理和传输办公所需的资料，改进工作环境。原始数据存储、电子转账和管理的电子商务信息是办公自动化系统的基本任务。

系统集成中心把智能化建筑的各个要素作为核心，利用结构化的综合布线系统和计算机网络技术，将语言、图像和数据信号，经过统一的筹划，综合设计在一套结构化的综合布线系统中，并以建筑物内外的综合布线系统和公共通信网络为桥梁，以及协调各类系统和局域网之间的接口和协议，把分离的设备、功能和信息有机地连接成一个整体，从而构成一个完整的系统，使资源高度共享，管理实现高度集中。

综合布线系统是将各种不同组成部分构成一个有机的整体，采取模块化结构设计，支持语音、数据和图像的传输，满足智能建筑对信息传输的要求。综合布线系统由6个子系统组成，它们是工作区子系统、水平布线子系统、管理子系统、主干布线子系统、建筑群子系统和设备间子系统。

建筑智能化的目的就是应用现代4C技术构成智能建筑结构与系统，结合现代化的服务与管理方式，给人们提供一个安全、舒适的生活、学习与工作环境。

Keys to Exercises of the Text

Ⅰ.
1．C 2．B 3．D 4．A 5．D

Ⅱ.
1．T 2．F 3．F 4．T 5．T

Ⅲ.
1．Building Automation System（BAS）　　2．HVAC
3．to take the modular design of the structure　　4．Integrated Services Digital Network
5．系统集成中心把智能化建筑的各个要素作为核心。
6．智能化建筑已经成为21世纪建筑业的发展主流。

第11单元　结构化综合布线

结构化综合布线系统是一个智能建筑必不可少的子系统，是用来进行语音、数据、图像等信息传送的网络。一个设计良好的布线系统应该具有开放性、灵活性和可扩展性，并对其服务的设备有一定的独立性，能充分满足现代社会信息技术的高速发展，使用户的投资得到一个可靠的保证。

结构化综合布线系统已经成为当今建筑物的基本设施。布线系统是建筑物或建筑群内的信息传递的媒介。它不仅将数据通信设备、话音交换设备和其他信息管理系统彼此相连，而且还连接着智能楼宇的其他子系统。其灵活性、兼容性和可靠性已得到中国用户的认可，并已经广泛地应用到国家职能部门、银行、大型集团公司、房地产公司等行业。结构化综合布线系统为用户提供了理想的布线方式，并依靠其高品质的材料，一改传统的布线方法，为现代化的大厦、工厂真正成为智能型楼宇奠定了15年内不需改变通信线路的传输媒介基础。

综合布线系统由以下6个独立的子系统组成。

1）工作区子系统由终端设备连接到信息插座的连线组成。常用设备是计算机（PC、工作站、服务器、打印机等)、电话、传真机、复印机等设备。

2) 管理子系统由交连、互连接线板组成,方便管理员进行系统管理,以方便实现配线管理,其设计很完善,很容易追踪跳线,体积小,比传统配线箱节省50%空间。

3) 水平子系统实现信息插座和管理子系统间(跳线架)的连接,采用超五类及六类双绞线实现这种连接。

4) 主干子系统实现机房管理中心到各楼层管理子系统间的连接。从核心管理子系统到设备管理中心的连接,建议采用光纤线缆做垂直主干,至于备用垂直主干线(子系统)可以采用超六类双绞铜线,其系统传输率可高达1GB/s。

5) 设备子系统通过为不同设备配置相应的适配器,实现网络布线系统与设备的连接。这里的设备是指安装在设备管理中心的以太网交换机、程控交换机、路由器服务器和防火墙等。

6) 建筑群子系统实现建筑物之间的相互连接。

尽管综合网络布线系统的费用只占整个网络结构费用的1/10,但近70%的网络问题都是与低劣的布线技术和电缆部件有关。因此,一个设计和组织良好的网络综合布线系统能够在安装、维护和升级中,节省大量资金和人力,总体拥有成本(TCO)平均下降30%。所以,一个良好的综合网络布线系统能给我们的工作、生活带来极大的方便。

Keys to Exercises of the Text

Ⅱ.
1. structured cabling system
2. exchange equipment
3. information transmission
4. terminal device
5. information management system
6. information outlet
7. network cabling system
8. the Ethernet Switch

Ⅲ.
1. (结构化综合布线)系统是用来进行语音、数据、图像等信息传送的网络。
2. 工作区子系统由终端设备连接到信息插座的连线组成。
3. 它很容易追踪跳线,体积小,比传统配线箱节省50%空间。
4. 所以,一个良好的综合网络布线系统能给我们的工作、生活带来极大的方便。

阅读材料　　　　　　　　**家居布线解决方案**

预先布线一个智能的家不再是奢侈的事情,而是一种需要。普天家庭布线系统是建立家庭联系与娱乐网络的基础。这个系统的设计可以组织以及分配新家的网络。布线模块通过5类或6类双绞线来连接各个房间到达普天配线箱。这种功能可以在你的整个屋子中管理或分配语音、数据、音频、视频以及安全信号。

普天家居布线系统将家庭中各种网络、电话、有线电视、视频及监控等智能家居设备信号传输的物理线路,包括插座、线缆、家庭多媒体配线箱等相结合,进行统一管理。该系统是智能家居各个设备的信息传输的通道,是实现其各自功能的基础。

普天家居布线系统相当于综合布线系统的工作区到管理间的家庭版本,也包括工作区、传输区和管理区,是综合布线系统的一部分。所不同的是信息点的种类更为多样,除了语音信息点、数据信息点之外,还有有线电视信息点、家庭影院(背景音乐)信息点、安防或监控信息点等类型。

家居布线的标准提供一个可满足电信服务最低要求的通用布线系统,该等级可提供电话、有线电视和数据服务。等级一按照星形拓扑，采用非屏蔽双绞线连接。这里使用的非屏蔽双绞线必须满足或超过 EIA/TIA－568A 规定的 3 类电缆传输比特要求。另外，还需一根 75Ω 同轴电缆，并必须满足或超过 SCTE IPS-SP-001 的要求，以便传输有线电视信号。建议安装超 5 类非屏蔽双绞线（UTP），以方便未来能升级到等级二。具体配置为：

1）每户可引入 1 条超 5 类对绞电缆；同步（使得住宅用户内部整齐美观，便于施工和维护，方便用户）附设 1 条 75Ω 同轴电缆及相应的插座，引入家庭信息配线箱。

2）每户宜设置嵌入式壁龛式家庭信息配线箱。

3）每间卧室、书房、起居室、餐厅等均应设置 1 个信息插座（用于数据或语音通信）和 1 个有线电视插座；主卫生间等其他房间还应设置用于电话的信息插座。

4）每个信息插座或有线电视插座至家庭信息配线箱，各敷设 1 条 5 类 4 对对绞电缆或 1 条 75Ω 同轴电缆，组建成星形网络。

5）如将安防系统接入家庭信息配线箱，应根据各种安防系统的要求，敷设相应线缆并配置相应终端设备。

6）家庭信息配线箱的配置应一次到位，满足远期的需要。

Keys to Exercises of the Reading

Ⅱ.

1. 普天家庭布线系统是建立家庭联系与娱乐网络的基础。
2. 普天家居布线系统相当于综合布线系统的工作区到管理间的家庭版本。
3. 等级一按照星形拓扑，采用非屏蔽双绞线连接。

第 12 单元　办公自动化系统

办公自动化系统包含来文管理、发文管理、阅文管理、档案管理、个人信息管理、会议管理等模块，可以实现企事业单位无纸化办公，提高办公效率，节约内部成本，提高经济效益。

1. 生产及使用条件

使用单位有良好的硬件运行环境，有中央服务器，各部门配备计算机，百兆以上的网络环境。

2. 技术特性

1）通用性：研制开发办公自动化软件，除针对特定企业开发专用性的软件外，应以开发通用性的办公软件为目标，适合各类企业事业单位的办公需求。各用户只要根据本单位的具体实际情况进行相应的配置，就可达到本单位的办公需求，实现由通用到专用的转换。

2）自动化：将办公中需重复处理的工作实现自动化。可以建立代理，由管理员在服务器上拥有无限制的 Lotus Script 代理权限，运行相应的代理就可以实现自动化的工作。

定期进行统计：统计单位内部各位领导、各级组织、各位成员是否及时办公，收文、发文、签报等工作的及时办理、超时办理、积压未办理的统计信息。

定期发送会议通知：每周末发送下一星期的会议安排表，此会议安排表是自动形成的，与会人员每人都会受到相应的会议安排。自动查询与会人员的空闲时间。安排会议时需要了解与会人员是否有空闲时间，与会人员繁忙时可另安排时间或重新安排人选。

自动删除过期文档：当用户工作很多时会积压大量过期文档，手工删除会增加许多工作量，设置定期删除代理，可以将过期文档自动删除。

自动催办：当文件或某工作在某单位积压时间超过规定期限，系统自动发出催办通知，催促该人员及时办理。

外出代理：当某位员工外出，应设置外出信息，这样其他人给此人发邮件或文件时，系统自动通知对方此人已外出，要求对方是否继续发送还是改换人选。

自动记录文件流转过程：收文或发文的所有过程与步骤，自动记录并显示，使文件流程清晰可见。

实现分组发送：设置单位内部的不同群组，发送邮件，发送会议通知，发送收文与发文时选择相应群组，就可由系统自动给群组内所有人员发送，而不必每个人都选择一次。当发送的对象很多时，此方式将大大降低发送人员的工作量。

系统自动提示有新的工作：当员工正在进行其他工作时，来了新的邮件或待办文件，系统可自动实现视觉与声音的提示，通知该员工及时办理。

自动确认邮件是否成功发送，对于重要的文件，此项功能十分关键。

3）安全性：为了保证信息交流过程中对重要信息的保护，设置不同级别的安全性控制，以达到符合用户要求的安全设置。

中国进入世界贸易组织后，中国企业将更多地面对世界的竞争，这样不仅要求企业加强管理，提高生产技术，还要求企业提高内部的办事效率，降低内部交易成本。利用网络办公平台可以实现无纸化办公，迅速传递大量信息，实现协同工作。

Keys to Exercises of the Text

Ⅱ.
1. T   2. F   3. F   4. T

Ⅲ.
1. network management   2. non‐paper office   3. the central server.
4. meeting management   5. personal information   6. technical traits
7. Besides being a mechanician he is a great poet.
8. The second considers that, as a developing country, China has done far more than its fair share.

Ⅳ.
办公自动化系统应实现的功能如下。企业的日常管理应包括企业的会议管理和公告管理。用户可以通过这一模块查看企业发布的会议、公告，并设置添加新公告和会议的功能。

第13单元 安全防范系统

随着时代的不断进步，科技的迅速发展，人们对居住、生活和工作环境的安全性要求也越来越高，迫切需要一个安全可靠、能预防罪犯的安全防范系统来保护人们的生命财产安全。智能化安防系统正是为解决人们的安全顾虑而产生的。

数字闭路电视监控系统和防盗报警系统是安全防范技术体系中的一个重要组成部分，是智能系统中不可少的子系统之一，是一种先进的、防范能力极强的综合系统。对于人们不便肉眼观测到的场合，它可以通过遥控摄像机及其辅助设备直接观看被监视场所的情况，一目了然；它提供了实时、形象、经济的监控手段。同时，它可以把被监视场所的图像部分或全部地记录下来，这样就为日后对某些事件的处理提供了方便条件及重要依据。数字闭路电视监控系统还可以与防盗报警等其他安全技术防范系统联动运行，使防范能力更加强大。因此，它在工业、小区、酒店及其他领域得到了广泛的应用。

由安全防范技术的器材、设备组成的系统能对入侵者做到快速反应，并及时发现或抓获罪犯，对犯罪分子有强大的威慑作用。而安全防范技术又能及时发现事故的隐患，预防破坏，减少事故或预防灾难，所以它是安全保卫工作中很重要的预防手段。尤其在现代化技术高度发展的今天，犯罪更趋智能化，手段更隐蔽，加强安防技术的现代化就显得更为重要。

一般所说的安全防范系统包括闭路监控系统、楼宇对讲系统、防盗报警系统、智能家居系统、出入口控制系统、消防报警系统等。

闭路监控系统是安防系统最重要的子系统，通过在现场布置一定数量的摄像机，进行安全监视和管理监视。

楼宇对讲系统是一种针对现代物业要求设计的保安系统，可集室内安防报警系统、出入口控制为一体，实现呼叫、对讲、视频监视、安防报警、开锁、三方通话等多种功能，达到智能化物业管理目的。

防盗报警系统通过在现场布置适量的移动探测器、红外探测器、门磁探测器、玻璃破碎探测器、烟雾探测器、紧急按键等各种探测器，当有非法入侵或任何异常情况时，都会发出报警信号，从而有效地保护人们的生命财产安全。

出入口控制系统通常采用现代电子与信息技术，在出入口对人和物这两类目标的进出进行放行、拒绝、记录和报警等操作，包括门禁控制、停车场管理等系统。

智能家居系统是安防系统的一个重要组成部分，通过相应的探测器和通信主机，将家居报警与远程控制相结合，不但具备防盗、监控、报警功能，还可以远程通过电话或手机遥控家中的任何一种电器设备的开关。

安防与智能是社会信息化的必然趋势。在高科技犯罪案例越来越多的今天，安全防范系统表现出了安全卫士的中坚作用，如闭路电视监控系统、周界防范系统、防盗报警系统、楼宇对讲系统、智能停车场管理系统、公共广播系统等。统计表明，安装了安防系统的建筑物案发率降低了约80%。

Keys to Exercises of the Text

Ⅱ.

1．T 2．F 3．T 4．T

Ⅲ.

1．security technology system
2．digital closed – circuit television monitoring system
3．burglar alarm system
4．communication host
5．building intercom system
6．intelligent home security system
7．entrance and exit control system
8．remote control

Ⅳ.

1．随着科学技术的飞速发展，人们对居住、生活和工作环境的安全需求也越来越高。
2．闭路监控系统通过在现场布置一定数量的摄像机进行安全监控和监控管理。
3．安防与智能是社会信息化的必然趋势。
4．统计表明，安装了楼宇安防系统使犯罪率降低80%左右。

阅读材料　　　　　　　　　　　　**智能化城市**

智能化城市着重于城市化发展、城市可持续增长和城镇居民的核心需求。智能化城市需要有效地融合先进的信息技术和先进的操作服务理念。（如图13-1所示）。智能化城市将通过实时收集并存储大量的城市信息资源来创建其IT基础设施，并且通过数据互联和互用性、交换、分享和合并应用程序，为生成和实现与城市管理操作相关的决策以及配置和管理创新公共服务创建一个平台，提供一种方便、高效、灵活的工具，从而实现和谐发展更安全、更环保、更高效、更方便的智能化城市的终极目标。

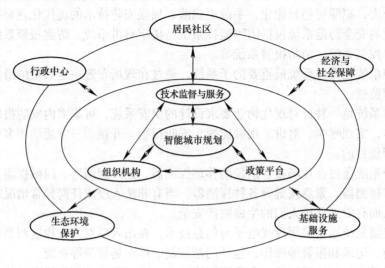

图 13-1　智能化城市

每一个智能城市都应该参照不同城市使用如下这些技术的例子。

1. 倡议开放数据和编程竞赛活动
2. 停车应用程序
3. 让用户采用城市属性的应用程序
4. 高科技垃圾管理系统
5. 全数字、易于使用的停车支付系统
6. 城市向导应用程序
7. 城市触摸屏
8. 覆盖地铁和火车的无线区域网络
9. 可持续、节能住宅和商业地产
10. 动态显示实时信息亭
11. 基于应用程序或社会媒体的应急预警和危机应对系统
12. 使用实时数据来监控和阻止犯罪的警力
13. 更多的公共交通、高速火车和快速公交系统
14. 犯罪高发区的有机发光二极管灯和监控
15. 类似太阳能的充电站
16. 覆盖着太阳能电池板和花园的屋顶
17. 自行车共享项目
18. 共享的经济，而不是购买经济
19. 家庭和企业中的智能气候控制系统
20. 重新规划交通路线的应用程序的广泛应用
21. 水循环系统
22. 广泛征求群众意见的城市规划
23. 为所有市民接入宽带互联网

24. 移动支付
25. 拼车应用程序

Keys to Exercises of the reading

Ⅱ.
1. 智能化城市需要先进的信息技术和先进的操作与服务理念的有效结合。
2. 智能化城市将通过实时收集并存储大量的城市信息资源来创建其IT基础设施。
3. 它将创建一个平台，这个平台提供一种方便、高效、灵活的工具用于生成和实现与城市管理和操作相关的决策。

第3章 自动化控制技术

第14单元 控制工程介绍

近年来，自动控制领域取得了相当大的进步，其实这种技术历史悠久，可以追溯到1790年，当时瓦特就发明了离心式调速器来控制蒸汽发动机的转速。他发现在一些应用中有必要保持发动机转速不随负荷扭矩的变化而变化。但实际上，当增加负载时速度就会减慢或者去除负载后速度就会增加。

在一个简单的离心调速系统中，发动机转速的变化被检测并用来控制进入发动机的蒸汽压力。在稳定条件下，因为离心力的作用和蒸汽阀的开度足够维持发动机转速所要求的水平，瞬间与金属摆球的重量平衡。当额外的负荷扭矩加到发动机上，发动机的转速会下降，离心力减少，金属球将轻微下降，其高度控制蒸汽阀的开度，当开度变大时将更多的蒸汽压力加到发动机上，这样转速就要上升，抵销了原先转速下降的倾向。如果额外的负载去除后，将发生相反的过程，金属球要轻微上升，蒸汽阀倾向于关闭，抵消了一些转速增加的倾向。

很明显，如果没有这种调节器，速度就会降到底，然而，经过一个合适的调节系统，转速下降会很少。随着高灵敏度的转速控制系统的出现，也会产生一些人们不希望出现的新问题，即控制量紧随被控制量（转速），从而在稳定的转速附近发生不规则的振荡和摆动。所有这种系统的真正问题是预防过度振荡的同时产生良好的调节作用。调节作用被定义为负载条件下被控制量的数值相对空载条件下被控量数值的变化百分比。各种调节器构成了一个重要的、完整的控制系统，它们通常能够保证各自所控对象（对应）的物理量（如速度、电压、液面高度、湿度等）在负载变化时保持恒定。一个好的调节器具有很少的调节量。

1914～1918年的战争，促使军队工程师意识到为了赢得战争胜利需要准确而迅速地使重型装备（如船只和枪炮）定位。在20世纪20年代早期，美国的N. Minorsky从事了这项关于轮船的自动驾驶和在船甲板上自动配置枪的杰出工作。在1934年术语"随动系统"（来源于拉丁文）第一次出现是在H. L. Hazen. 的文学作品中。他将伺服机构定义为一个能量放大装置，在该装置中，用以驱动输出量的放大元件由来自于伺服机构的输入和输出之间的偏差来控制。这种定义能广泛应用于多种多样的反馈控制系统。最近有人认为术语"随动系统"和"伺服系统"受反馈控制系统的机械位移变量限制。

在过去的30年间，控制工程通用领域中的一个极其重要的部分"自动控制"已经出现，而且被广泛应用于诸如化工、食品加工、金属加工等各种各样的大规模的工业过程控制中。在发展的最初阶段，很难想象过程控制理论与随动系统和调节器理论密切相关。甚至到现在，过程控制系统的完善设计实际上不可能归功于我们对过程动力学的那些有限的理解。

本书介绍的大部分理论都以随动系统和调节器为例阐述分析方法。

Keys to exercises of the Text

Ⅱ.

1. at 2. invented 3. increased 4. detected

Ⅲ.

1. It is obviously that while unloading, the speed of engine would rise considerably.

2. Due to the centrifugal force, the metal spheres are just sufficient to maintain the engine speed at the required level.

3. In the initial stages of research, it was scarcely realized that the theory of control is of importance.

4. Regulation is defined as the percentage change in controlled quantity on load relative to the value of the controlled under condition of zero load.

第15单元　可编程序逻辑控制器

可编程序逻辑控制器，PLC，或者称可编程序控制器，它是一种小型计算机，可用于自动化生产，比如控制工厂的流水线等。可编程序控制器通常使用了微处理器进行控制。其程序可控制复杂的加工工序，并由工程师编写。程序通常存储在随机存储器和（或）电可擦写可编程只读存储器中。

PLC 的基本结构如图 15-1 所示。

从图中可见，PLC 有 4 个主要的部件：程序存储器、数据存储器、输出设备和输入设备。程序存储器用来存储逻辑控制序列指令。开关状态，联锁状态、数据的初始值等其他工作数据存储在数据存储器中。

与其他计算机比较，PLC 最大的特点是有专用的输入/输出装置。这些装置将 PLC 与传感器和执行机构相连接。PLC 能读取限位开关、温度指示器和复杂定位系统的位置信息，有些甚至使用了机器视觉。执行机构方面，PLC 能驱动任何一种电动机、气压或液压缸、薄膜装置、电磁继电器或螺线管。输入/输出装置可以是安装在 PLC 内部，或者通过外部 I/O 模块与专用的计算机网络相连。可编程序控制器是作为要使用成百上千个继电器和凸轮的早期的自动控制系统的低成本替代品被发明出来的。一台 PLC 经常可以通过编程来替代数千个继电器。可编程序控制器最初在自动化的制造工业中使用，通过对软件的修改可以取代硬连线控制面板的重新连线。

可编程序控制器经过数年的发展，其功能包括典型的继电器控制，复杂的运动控制，过程控制，分布控制系统和复杂的网络技术。最早的 PLC 采用简单的梯形图来描述其逻辑程序，梯形图源自于电气连线图。电气技师们使用梯形逻辑图很容易找出原理图中的电路问题。这样大大减少了技术人员的忧虑。

如今，PLC 的功能已经非常可靠，但是可编程计算机仍然有其发展空间。依照 IEC 61131-3 标准，现在 PLC 可以使用结构化的编程语言和逻辑初等变换来进行编程。

某些可编程序控制器可使用一种图表式的编程符号，这种符号称为顺序功能图表。

然而，值得注意的是，PLC 不再是非常昂贵的产品（通常是数千美元），而成了一种普通的产品。如今，功能完善的 PLC 也只要数百美元。使机床实现自动化还有其他的方法，比如传统的基于微控制器的设计，但是这两者之间有一些差别。PLC 包含所有能直接操作大功率负载的输出功能，而微控制器还需要电气工程师设计专用的电源和电源模块等。同时微控制器设计的方式不具备 PLC 现场可编程的灵活性。那也正是 PLC 在生产线上运用的原因之一。PLC 控制系统是典型的高度可定制系统，所以与一次请一个设计人员做一个专门的设计比起来，一台 PLC 的成本要低一些。另一方面，在大批量生产中，因为其构件成本低廉，采用的可定制控制系统能很快收回成本。

Keys to exercises of the Text

Ⅱ.

1. F 2. F 3. T 4. T 5. T

Ⅲ.

1. The PLC usually uses a microprocessor.

2. PLCs read limit switches, temperature indicators and the positions of complex positioning systems.

3. The earliestPLCs expressed all decision making logic in simple ladder logic.

4. PLCs contain everything needed to handle high power loads right out of the box.

阅读材料   **PLC 编程**

最初 PLC 编程是采用与继电器逻辑原理图相同的技术（梯形逻辑图），这样电工、技术员和工程师都不需要学习计算机编程。但是这种方法被普遍接受了，成为今天 PLC 编程的最通用的技术。图 15-2 给出一个梯形逻辑图，为了说明这个图，我们可以想象左边的竖线为电源线，称为火线，右边的是零线。图中有两条通路，每一条通路上都有输入或输入的组合（两条水平线）和输出（圆形符号），如果输入的开闭组合正确，电流就可以从火线通过输入，流过输出最终到零线。输入可以来自传感器或开关。输出可以是 PLC 外接的一些器件，它控制如灯或电动机合上或断开。在第 1 条通路中连接一个常开和一个常闭，如果输入 A 是合上（则 A 动作，其常开触点成为闭合），B 是断开（则 B 不动作，其常闭触点仍为闭合），则电流会流过输出并使输出有效。而任何其他输入的组合（另 3 种分别为：A 合上，B 合上；A 断开，B 合上；A 断开，B 断开）都会使（第一条通路不通）输出 X 是断开的。

图 15-2 的第 2 条通路比较复杂，有多个可能的输入组合会使输出 Y 接通。在通路左部，上面一条如果 C 断开，D 接通电流可以流过；下面一条，如果 E 和 F 都接通电流也可以流过，这时电流流过半条通路；如果还有 G 或 H 接通则电流就流到输出端 Y 了。

PLC 编程还有其他方法，其中最早的技术之一是符号指令。这些指令可以直接从梯形逻辑图中导出，通过一个简单的编程终端输入到 PLC。图 15－3 是符号指令的一个例子，在这个例子中，可以从顶部到底部，每次读一条通路得到符号指令。第 0 行是对输入端 00001 的指令 LDN（输入加载并取反）。这条指令会检查输入端 00001，如果是断开就输给 PLC 一个 1（或真），如果输入端接通就输给 PLC 一个 0（或假）。下一条用 LD（输入加载）语句来检测输入端 00002，如果输入端是断开就记为 0，如果输入为通电则记为 1。AND 语句是把上两条语句的值相"与"，如果都为真则结果为 1，否则结果为 0。对 00003 和 00004 的输入处理过程相同，第 5 行 AND 指令把后两个 LD 指令的输出值相"与"得到一个输出值。OR 指令对保留的两个值进行"或"运算，如果任何一个为 1，输出为 1，否则结果为 0。最后的指令是 ST（存储输出），会把最终的输出值存储起来，如果是 1 则输出端为通电，如果是 0 则输出端为断开。

图 15-3 的梯形逻辑程序与代码程序等效，即使你是用梯形逻辑图编程的，也要把它转换成代码形式输给 PLC，PLC 才能执行。

Keys to Exercises of the reading.

Ⅱ.

1. 最初的 PLC 编程使电工、技术员和工程师都不需要学习计算机编程。
2. 输出可以是 PLC 外接的一些器件，它控制如灯或电动机的合上或断开。
3. 符号指令是最早的技术之一。
4. 这些指令可以直接从梯形图中导出，通过一条简单的编程引脚输入到 PLC。

第 16 单元　电子检测设备

如图 16-1 所示，电子检测设备一般由 3 部分组成。

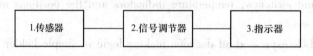

图 16-1　电子检测设备的组件

传感器将非电信号转换成电信号，因而，只有当所测量的电量是非电信号（例如压力）时才需要传感器。

用信号调节器处理输入的电信号，将其变为适用于指示装置的信号。信号可能需要放大到足够的振幅以使指示装置产生明显的变化。其他类型的信号调节器也可能是分压器（它能够减少输入到指示装置的信号量），或整流器、滤波器、斩波器等整波电路。

常用的指示装置一般是偏转型仪表，如电压表、电流表、欧姆表等通用仪表。

电子检测设备可以用来测量电流、电压、电阻、温度、声级、压力及其他物理量。然而，尽管在指示仪表的刻度盘上显示出了度量单位，但由于电流的流动，指针的显示是有偏差的。

如果处理不当或使用不当，最好的电子仪表也会显示不准确的结果。只要遵守一些基本原则，一般就可以保证仪表显示的测量数据是可以接受的。

多数仪表都是精密敏感的装置，使用时应十分小心。使用仪表之前，应当完全熟悉其操作规程。仪表的操作和使用说明书提供了详细资料，每一件新购进的仪表都配有说明书。电子实验室应将说明书存档以便查阅。如果对仪表的操作、规格、性能、限制条件等不完全熟悉，在使用仪表之前要阅读使用说明书。

应当选择其精确度能满足要求的仪表。虽然精确度和分辨率越高越好，但一般来说仪表的价格却是与这些性能直接相关的。

一旦选定了所使用的仪表，就应当查看是否有任何明显的外表损伤，如旋钮是否松动，外壳是否损坏，指针是否弯曲，手柄是否摇晃，测试引线是否受损等。如果仪表有内置电池电源，在使用前要检查电池的状况。为此许多仪表都有"电池检查"位置。当必须更换电池时，要确保所更换的电池及其安装均正确无误。

将仪表接入电路之前，要确保功能开关位置正确，范围选择开关也位于正确的范围之内。如果在选择范围上出现疑问，在将仪表接入电路之前，应将仪表调到最大范围；然后将仪表向较低的范围调整，直到读数接近中间刻度为止。为了使所测试设备能获取最准确的数据，还要考虑其他的因素，如电路负载、阻抗匹配及频率响应等。

Keys to exercises of the Text

Ⅱ.

1. F　2. F　3. T　4. T　5. F

Ⅲ.

1. The signal may need to be amplified until it is of sufficient amplitude to cause any appreciable change at the indicating device.

2. Before using an instrument one should be thoroughly familiar with its operation.

3. You should select an instrument to provide the degree of accuracy required.

4. Many instruments have a "battery check" position.

Ⅳ.

该公司主要从事电气和电子仪器、电子测量仪器制造、技术开发、技术服务和网络配电自动化业务。

第 17 单元　适应性控制系统

适应性控制系统是一种自动调整其参数以补偿过程特性的相应变化的系统。简单地说，该系统可以"适应"过程的需要。当然，必须有一些作为适应程序依据的准则。为被调量规定一个数值（即设定值）是不够的，因为要满足这一指标，不仅需要适应性控制，还必须另外规定被调量的某种"目标函数"。这是一个决定需要何种特殊形式的适应性控制的函数。

一个给定过程的目标函数可能是被调量的衰减度。因而，实质上有两个回路，一个回路靠被调量操作，另一个则依赖其衰减度。由于衰减度标志着回路动态增益，因此这种系统被称为动态适应性系统。

也有可能为一个过程规定一个静态增益的目标函数。为这种指标而设计的控制系统就是静态适应性系统。

实际上，这两种系统之间几乎不存在相似，以致在同一名称"适应性"之下，它们的分类已经引起了许多混淆。

要指出的是，第二个区别并非是目标函数，而是关于如何实现适应性控制的机制问题。如果对于过程有充分的了解，使得参数的调整能够与那些引起过程性质变化的变量有关系，那么适应性控制就可以程序化了。然而，如果必须根据目标函数的测量值来调整参数，则要利用反馈回路实现适应性控制，这种系统称为自适应系统。

1. 动态适应性系统

动态适应性系统的主要功能是给控制回路一个始终如一的稳定度。因此，动态回路增益就是被调量的目标函数，其数值要予以规定。

最容易变化的过程特性是增益。在某些情况下，静态增益会发生变化，这种情况通常称为非线性。另外一些过程表现出可变周期，这就对动态增益产生了影响。但是不论回路的稳定性受到哪种机制的影响，都能通过适当调整调节器的增益来恢复稳定性（这里假定所希望的衰减度是可以达到的，这就排除了极限环的情况。）

有关可变过程增益的许多情况已经叙述过了。通常，为了补偿这些变化，采取的办法是在控制系统中引入一个经过选择的非线性函数。例如，调节阀的特性是考虑到这个目的通常选择；但是用这种方法所作的补偿可能由于下列原因而不能满足。

1）引起增益变化的根源位于回路以外，从而不能从调节器的输入或输出识别这些变化。

2）所需要的补偿是几个变量的综合函数。

3）过程增益随时间而变化。

2. 静态适应性问题

凡动态适应性系统都是控制回路的动态增益，那么，与它相对应的静态适应系统就寻求不变的静态过程增益。当然，这意味着静态过程增益是变化的，而且有一个特定值是所期望的。

以控制燃烧系统为例，要获得最高燃烧效率，应调节其燃油、空气比。过量燃油或过量空气都会降低燃烧效率。真正被调量是燃烧效率，而真正的控制量是燃油 - 空气比。本例中，期望的静态增益是 $dc/dm = 0$。该系统应在这样的控制点上运行，即在这一点上燃油 - 空气比无论是增大或减小，都会降低燃烧效率。这是静态适应性控制的一种特殊情况，称为"最佳化"，然而，也可以合理地规定一个非零增益。

满足目标函数的被调量数值是与该过程中主要条件相关的情况下，那么就能够容易地为适应性控制编制出程序。例如，在各种空气流量和温度情况下的最佳燃油 - 空气比可以是已知的。因此，用改变控制器设定值的方法来设计控制系统，使燃油 - 空气比适应于空气流量与温度的变化，作为动态适应系统例子中的一个流量函数。

Keys to exercises of the Text

Ⅱ.
1. parameters 2. meet 3. determines 4. resemblance
5. self adaptive system 6. control loop 7. control valve
8. fuel – air ratio 9. adapt

Ⅲ.
1. 还必须另外规定被调量的某种"目标函数"。
2. 在某些情况下，静态增益会发生变化，这种情况通常称为非线性。
3. 然而，如果必须根据目标函数的测量值来调整参数，则要利用反馈回路实现适应性控制。
4. The steady – state process gain is variable and that one particular value is most desirable.
5. Where the value of the manipulated variable which satisfies the objective function is known relative to conditions prevailing within the process, the adaption may be easily programmed.

阅读材料　　　　　　　　控制系统组件

众所周知，一个闭环控制系统包括以下 3 个基本组件。

1）误差检测器。该装置接收低功率输入信号和具有不同物理性质的输出信号，将他们转换成一个常见物理量用来做减法，并且输出一个低功率的具有正确物理性质的误差信号来启动控制器。误差检测器通常包含传感器，用以实现将一种物理形式的信号转换为其他形式。

2）控制器。这是一个放大器，它可以接收低功率误差信号，同时加上来自外部电源的功率。接着一个大小可控的功率便提供给了输出元件。

3）输出元器件。它可以给负载提供合适的功率，这个功率与从控制器接收的信号一致。

Keys to Exercises of Reading

1. three 2. an amplifier 3. transducers 4. controller

第 18 单元　自动控制系统

自动控制系统是一个预先设定的闭环控制系统。一个自动控制系统有两个过程变量：控制量与被控量。被控量要保持为一个指定的值或在指定的范围内。例如，在水箱水位控制系统中，储水池的水位是被控量。控制量是这样一种过程变量，它是由控制系统控制，用来使被控量能保持在一个理想值或指定范围内。在前例中，水注入池中的速度便是一个控制量。

1. 自动控制系统的功能

任意一个自动控制系统都包括 4 个基本功能：测量、比较、计算、修正。在水箱水位控制系统中，水位感应器测量水池中的水位。并将表示水位高低的信号传递给控制装置，控制器将该信号与所指定的信号进行比较。然后，水位控制装置计算出供水阀要打开多少才能减小实际水位和指定水位之间的差距。

2. 自动控制系统的基本组成

自动控制系统由 3 种基本元器件构成：测量元器件、误差检测元器件和末控制元器件。这些元器件之间的联系以及它们所执行的功能，如图 18-1 所示。测量元器件执行测量的功能，即传送和计算被控量。误差检测元器件先将被控量的值与期待值作比较，如果实际值和期待值间有偏差，便给出一个信号。末控元器件通过纠正系统的控制量对所给的误差信号进行响应。

3. 反馈控制

自动控制器是一个误差检测，自动修正的装置。它从控制过程中获得信号，再将信号反馈给控制过程。因此，闭环控制即是人们所说的反馈控制。

图 18-2 中的方框图所示的是反馈控制系统的基本组成。这些基本元素在功能上的联系是显而易见的。有一个很重要的地方请大家记住，方框图中显示的是控制信号的流程，而不是能量在系统或控制过程中的流向。

以下是与闭环控制方框图相关的几个术语。这些元素也可称为"控制器"。反馈环节用来确定反馈信号和控制输出之间的功能关系。参考点是一个外部信号，将其提供给控制系统的比较点，用来生成使控制设备产生指定动作的信号。能反映被控量的期待值的信号也称为"设定值"。控制输出量通常是数值或受控设备的状态。这个信号表示的是被控量。反馈信号传送到比较点，并且在此与输入的参考信号进行代数相加以得到控制信号。这个控制信号表示的是控制环的状态，同时它的值是参考输入信号和反馈信号的代数和。该信号也称为"误差信号"。控制量是用来维持设备输出（被控量）所期待的值。而扰动量是系统不需要的输入信号，它将扰乱设备控制输出值。

4. 控制系统的稳定性

之前所描述的所有控制方式都能在扰动之后使过程值返回一个稳定值，这个特点称为"稳定性"。

控制环可能稳定也可能不稳定。不稳定性是由系统处理过程中的时间滞后和系统固有的时滞引起的。这将导致被控量变化响应迟钝。从而使被控量不断地在设定值周围变化。人们用振荡来描述这种特征。在控制环中有 3 种振荡方式：衰减振荡、等幅振荡和发散振荡。每一种振荡的波形图如图 18-3 所示。图 18-3a 所示的是衰减振荡。这种振荡振幅是衰减的，并且最终振荡会停止以阻止被控量的变化。这是自动控制系统中所期望得到的一种方式。图 18-3b 所示的是等幅振荡。这种情况下控制器的动作维持着被控量的振荡。被控量永远也不能达到稳定状态，因此这种情形不是人们所期待的。图 18-3c 所示是发散振荡。这种状态下，控制系统不仅要维持振荡，而且振幅不断增大。控制元件将会达到它们的调整极限，而导致系统失去控制。

Keys to Exercises of the Text

Ⅱ.
1. the storage tank level, the process variable
2. sending, evaluating
3. decreasing amplitude, constant amplitude, increasing amplitude
4. feedback control
5. controller

Ⅲ.
1. 水位传感器将表示水位高低的信号传递给控制装置，控制器将该信号与所指定的信号进行比较。
2. 这将导致被控量变化响应迟钝。
3. 不稳定性是由系统处理过程中的时间滞后和系统固有的时滞引起的。
4. The control element has reached its full travel limits and causes the process to go out of control.
5. It takes a signal from the process and feeds it back into the process.

第 19 单元　自动控制的应用

虽然自动控制的应用范围实际上是无限的，但是我们的讨论仅限于现代工业中常见的几个例子。

1. 伺服机构

虽然伺服机构本身并不是一种控制的应用，但是这种装置在自动控制中却是常用的。伺服机构，或简称为"伺服"，是一种闭环控制系统，其中的被控变量是机械位置或机械运动。该机构的设计使得输出能迅速而精确地响应输入信号的变化。因此，我们能把伺服机构想象为一种随动装置。

另一种控制输出变化率或输出速度的伺服机构称为速率或速度伺服机构。

2. 过程控制

过程控制是用来表示制造过程中多变量控制的一个术语。化工厂、炼油厂、食品加工厂、鼓风炉、轧钢机都是自动控制用于生产过程的例子。过程控制就是把有关诸如温度、压力、流量、液位、黏度、密度、成分等这样一些过程变量控制为预期值。

现在过程控制方面的许多工作都包含推广使用数字计算机，以实现过程变量的直接数字控制（DDC）。在直接数字控制中，计算机是根据设定点的数值和过程变量的测量值算出操纵变量值的。计算机的判定结果直接送给过程中的数字启动器。由于计算机兼有了模拟控制器的作用，所以就不再需要这些常规的控制器了。

3. 发电

电力工业首先关系到能量的转换与分配。发电量可能超过几百兆瓦的现代化大型电厂需要复杂的控制系统来负责处理大量相互关系复杂的变量，并提供最佳的发电量。发电厂的控制一般被认为是一种过程控制的应用，而且通常有多达 100 个操纵变量受计算机控制。

自动控制已广泛地用于电力分配。电力系统通常由几个发电厂组成。当负载波动时，电力的生产与传输要受到控制，使该系统达到运行的最低要求。此外，大多数的大型电力系统都是相互联系的,而且两系统之间的电力流动也受到控制。

4. 数字控制

有许多种加工工序,如镗孔、钻孔、铣削和焊接都必须以很高的精度重复进行。数字控制是这样一个系统,该系统使用的是称为程序的预定指令来控制一系列运行。完成这些预期工序的指令被编成代码,并且存储在如穿孔纸带、磁带或穿孔卡片等某种介质上。这些指令通常以数字形式存储,故称为数字控制。指令标识出使用工具、加工方法(如切削速度)及工具运动的轨迹(位置、方向、速度等)等参数。

5. 运输

为了向现代化城市的各地区提供大量的运输系统,需要大型、复杂的控制系统。目前正在运行的几条自动运输系统中有每隔几分钟的高速火车。要保持稳定的火车流量及提供舒适的加速和停站时的制动,就需要自动控制。

飞机的飞行控制是在运输领域中的另一项重要应用。由于系统参数的范围广泛以及控制之间的相互影响,飞行控制已被证明为最复杂的控制应用之一。飞机控制系统实质上常常是自适应的,即其操纵本身要适应于周围环境。例如,一架飞机的性能在低空和高空可能是根本不同的,所以控制系统就必须作为飞行高度的函数进行修正。

船舶转向和颠簸稳定控制与飞行控制相似,但是一般需要更大的功率和较低的响应速度。

Keys to Exercises of the Text

Ⅱ.

1. h 2. d 3. f 4. b 5. g 6. c 7. e 8. a

Ⅲ.

1. extensively 2. distribution 3. commonly 4. generating 5. minimum 6. interconnected

Ⅳ.

1. 伺服机构是一个闭环控制系统,系统中的控制变量是机械位置和运动。
2. 船舶转向的控制与飞行控制相似。
3. 电力系统通常由许多发电厂组成。
4. 自动控制已广泛用于电力的分配。

阅读材料　　　　　　　　**变频调速(驱动)系统**

变频调速系统是通过控制供给电动机的频率来控制交流电动机的旋转速度的。可变频率驱动是一种调速驱动的形式。变频驱动也称为可调频率驱动(AFD)、变速驱动器、AC驱动器、微驱动器或逆变驱动。因为随着频率变化电压发生变化,有时也称为变压变频(VVVF)驱动。

一个变频驱动系统一般由交流电动机、变频控制器和操作界面组成(图19-1)。

变频调速系统中所用的电动机一般是三相异步(感应)电动机,有时也用单相异步电动机,但三相电动机用得比较多。在有些场合各种同步电动机有优势,但对大部分应用来说异步电动机更适用,是更加经济的选择。一般选用主电路电压下按固定速度运行的电动机,但在标准电动机的设计上作一定的改进,可使之工作可靠性更高,并有较好的频率调速性能。

变频驱动控制是固态电子技术的功率变换设备。通常的设计是先把输入的交流电通过桥式整流转换成直流电（作为中间量），然后再把直流电用一个逆变开关电路转换成准正弦交流电源（图 19-2）。一般用二极管构成三相桥式整流器，有时也用可控制整流器。

操作者用操作界面来起动或停止电动机，调节电动机的速度。附加的操作控制功能包括使电动机反转，在手动速度调节和根据外部处理控制信号自动控制之间切换。操作界面通常有图形界面显示，指示灯和仪表用来表示（系统）驱动运行的信息。

Keys to Exercises of the Reading

Ⅰ.
1. 变频驱动系统　　　　2. 交流电　　　　　3. 旋转速度
4. 可调速度；变速　　　5. 变频　　　　　　6. 逆变驱动
7. 变压变频　　　　　　8. 三相电动机　　　9. 单相电动机

Ⅱ.
1. A variable-frequency drive (VFD) is a system for controlling the rotational speed of an alternating current (AC) electric motor by controlling the frequency of the electrical power supplied to the motor.

2. A variable frequency drive system generally consists of an AC motor, a VFD controller and an operator interface.

3. Variable frequency drive controllers are solid state electronic power conversion devices.

4. The operator interface provides a means for an operator to start and stop the motor and adjust the operating speed.

5. The operator interface often includes an alphanumeric display and/or indication lights and meters to provide information about the operation of the drive.

第4章　工业4.0与中国品牌制造

第20单元　工业4.0介绍

工业4.0起源于制造强国德国。然后这个概念性的想法已经被很多工业国家广泛接受和采用，这些国家包括欧盟内部的成员国，更远到中国、印度和其他亚洲国家。工业4.0这个名字指的是第四次工业革命，前三次工业革命分别通过机械化、电力和IT技术实现了。

第四次工业革命，也因此称为4.0，将通过物联网技术和互联网服务与制造环境的集成而发生。以前的工业革命带来的好处已经既成事实，而我们有机会积极地引导第四次工业革命的发展路径来改变我们的世界。

工业4.0的远景是，将来的工业企业将建立全球网络来连接他们的机器、工厂和库房设施来形成一个信息物理系统，这个系统将通过分享信息来引发动作，从而智能地连接各个部分并进行相互控制。这个信息物理系统将呈现智能工厂、智能机器、智能仓储设施和智能供应链等各种形式。这将带来工业过程的进步，在这个过程中，生产制造作为一个整体，这个改进将通过工程、材料使用、供应链和产品生命周期管理的提升来实现。这也是我们所称的水平价值链。对水平价值链的展望是，工业4.0将在水平价值链的每个阶段进行深度集成，从而使工业生产过程发生巨大的变化。

这个远景的中心是智能工厂，它将使生产方式发生改变，这个改变基于智能机器也基于智能产品。不仅智能工厂等信息物理系统是智能的，被组装的产品也包含嵌入式智能，这样的话它们就能够在生产过程中的每个阶段被标识和定位。缩微化的无线射频识别标签使产品变得智能并能够知道它们是什么，什么时候被生产，更为关键的是，它们目前的状况是什么和达到期望的状态还需要什么步骤。

这就需要智能产品能够知道它们自己的历史和把它们变成一个完整的产品还需要的工序。这些工业生产制造的知识将嵌入到产品中且允许它们提供生产工艺的替代路线。比如，当智能产品知道了它当前的状态和成为完整产品所需要的后续生产过程时，智能产品将能够指挥输送带，这也是它所要跟随的生产线。然后，我们将看看这个工作实际上是怎样运行的。

然而，现在我们需要看看工业 4.0 远景的另一个关键因素，那就是价值链中垂直制造过程的集成。（人们）所持的远景是：嵌入式水平系统与垂直业务过程（销售、物流、财务和它们之中的一些其他项目）和相关的 IT 系统的集成。它们将使得智能工厂能够对从供应链到服务和产品生命周期管理的整个制造过程进行首尾相连的控制。这个运营技术和信息技术的融合不是没有问题，我们在早些时候讨论工业互联网时就发现了这个问题。然而，在工业 4.0 系统里，这些独立存在的实体将扮演同一个角色。

智能工厂不仅仅是与大公司相关，实际上他们的灵活性使他们更适用于中小企业。比如，对水平制造过程和智能产品的控制使得我们更好地进行决策和进行动态过程控制，在能力和灵活性方面来适应最新的设计变更或者改变生产来满足顾客在产品设计方面的偏好。更进一步说，这个动态过程控制使小批量（生产）成为可能，而小批量（生产）还是盈利的且能够适应个性化订单。这些动态业务和工程过程使得创造价值的新方法和创新的商业模型成为可能。

总的来说，工业 4.0 需要信息物理系统集成在制造和物流过程中，同时在制造过程中引进物联网和互联网服务。这将带来新的方法来创造价值、商业模型和为下游的中小企业提供服务。

Keys to Exercises of the Text

Ⅱ.

1．f　2．d　3．a　4．b　5．g　6．e　7．h　8．c

Ⅲ.

referred; encompasses; benefit; optimize; via; digitalized; providing; communications

Ⅳ.

1．第一次工业革命通过使用水和蒸汽的力量实现了机械化。

2．第二次工业革命可以追溯到亨利·福特在 1913 年引进的组装线，它使产能大幅提升。

3．第三次工业革命是在 20 世纪 70 年代计算机导入到生产现场的结果，这引起了自动组装线的大量出现。

4．工业 4.0 的展望是"计算机－物理生产系统"——在这个系统中布满传感器的智能产品告诉机器它们将被怎样处理。

第 21 单元　中国高速铁路

"高铁"，被誉为是 20 世纪后期和 21 世纪最具革命性的交通工具。今天，"中国高铁"正是这种"最具革命性交通工具"的佼佼者。其实，中国高铁起步也就是从 2004 年年初才开始。那时，中国刚刚颁布其铁路史上第一个《中长期铁路网规划》。之后到现在，也就仅仅六年多时间，中国铁路实现了跨越式发展，中国铁路在让世人惊叹中昂首跨入了"高铁时代"。

2010年9月28日，中国自主研发的"和谐号-CRH380A"新一代高速动车组实验列车，以时速416.6km的超高速，在上海至杭州的高速铁路上创造了世界铁路运营史上的最高速度。"中国高铁"已经走到了世界高速铁路的最前沿，成为"中国速度"和"中国制造"最具标志性的范例。

按照中国规划的宏伟蓝图，到2020年，中国铁路营运总里程将再增加近一倍，突破12万km，其中高速铁路将超过1.6万km以上。届时，中国铁路将形成包括京－沪高铁、京－深－港高铁在内的"四纵四横"的高速客运骨架网。那时，北京到上海的旅行时间将由10小时大大缩短为不足4小时；北京至广州将由22小时锐减至6个半小时；北京至昆明将由最快38小时变为不超过8小时；甚至从北京到遥远的乌鲁木齐，也将由40小时缩短至仅仅11个小时。

正如美国《新闻周刊》所言：中国正在进行一场"铁路革命"，时速达350km以上的高铁让多年来以幅员辽阔为特色的中国正在"大大缩小"，并改变了这个国家的经济版图。

Keys to Exercises of the Text

Ⅱ.
1. T 2. T 3. F 4. T 5. T

Ⅲ.
1. is being promoted 2. around 3. are being urged 4. actively 5. to employ

阅读材料　　　　　　　**中国品牌来到拉斯维加斯国际电子产品展销会**

全球最大的电子产品展览会每年一月份在拉斯维加斯举行。超过1100多家的中国企业参加了2016年1月的内华达州拉斯维加斯消费电子展（CES），这比去年的550家增加了不少，约占3600多家参展商的1/3。

创新策略公司总裁蒂姆·巴加林在最近的科技时尚博客中发帖说："没错——中国人来了，他们计划尽其所能打乱传统消费电子生产商的阵营，并快速抢占他们的市场份额。"对于像华为技术有限公司和中兴通讯股份有限公司这样的顶级中国企业，CES仍是发布新产品的场所。对于像阿凡达机器人（AvatarMind）和HandScape这样的新选手来说，消费电子展可能使他们踏入国际市场的大门。

中国公司在展会上存在的另一个迹象是，中国企业占据了最佳地段的大量展位。当成千上万的参观者进入主厅时，首先映入眼帘的是美国半导体巨头——英特尔公司的展位。侧面相邻的展位被中国主要的电视机生产商海信有限公司和四川长虹电器有限公司占据。中国参展公司出现在每一个领域，从电视和电话到无人机和拥有3~4岁儿童身高的IPAL机器人。IPAL机器人由总部位于南京的阿凡达机器人科技有限公司开发，专为那些"有时感到孤独"或"希望有一个好朋友一起玩"的孩子开发。

它有两个版本：一个形似男孩，另一个形似女孩。IPAL可以讲故事，还可以一边晃动头部和手臂，一边唱《老麦克唐纳有个农场》（Old MacDonald Had a Farm）之类的儿歌；它还可以教汉语和英语。因为有一个检测手势的传感器，孩子们还可以与IPAL玩游戏。在它的胸口有一个六英寸的平板计算机来运行安卓操作系统，从而为MyPal（我的好朋友）开发应用程序。该公司表示，机器人很快就能教授孩子们"陌生人危险"这样的课程，教孩子们不要太接近陌生人。阿凡达机器人科技有限公司表示，它们正与位于康涅狄格州的一家公司合作，这家公司致力于开发针对自闭症儿童的机器人教育程序。

在电视厂商中，在其智能电视研发中投入巨资的海信公司在展会上展出了22个新机型，从720像素的小机型到4K曲面超高清电视（4K Ultra HD）。2014年海信的年营业额超过160亿美元，在全球电视机市场排名第三，比去年上升一位，在全球4K电视市场排名第二。海信也参与大型体育赞助活动，包括纳斯卡车赛和一级方程式赛车，未来它将更积极大胆地进行赞助活动。

中兴通讯作为中国最有名的电信设备制造商，也是中国领先的移动设备制造商，还以用得起的智能手机技术打入美国市场。它邀请前NBA球员拉里·约翰逊、拉里·南斯和阿多纳尔·福耶尔登台，还将NBA全国冠军赛的奖杯带到了展台上。

世界排名第1的个人计算机制造商联想，推出了世界上最薄的可折叠笔记本式计算机YOGA 900S，ThinkPad X1平板计算机，以及为企业培训提供虚拟教室的AirClass。

联想集团首席执行官杨元庆说，"越来越多的中国企业将成为全球知名品牌。这很难，但我认为随着成功案例的增多，这也将变得更容易。有一天，我们会看到中国品牌数量的大量增加，我们这样的公司已经铺平了道路。"

Keys to Exercises of the Reading

Ⅰ.

1. F 2. T 3. F 4. T 5. T

Ⅱ.

1. The first CES took place in New York City in June 1967. Since then, thousands of products have been announced at the yearly show, including many that have transformed our lives.

2. The International CES (Consumer Electronics Show) is the world's gathering place for all who thrive on the business of consumer technologies.

3. As an active angel investor there is no better place than CES to discover early stage companies in CE and related area and to learn the latest trends in technology.

4. Grow your global brand, increase your international presence, mingle with tech innovators, build your network and see the latest technologies that might just change the world.

第22单元　天空为无人机制造商尽显神通

中国世界领先的无人机行业逐渐占据了全球消费无人机市场的大半部分，而且在未来无人机技术的应用领域将更加广泛。

在1月份拉斯维加斯举办的国际消费电子产品展销会上，中国最大的商业无人机制造商大疆公司推出了两款消费无人机。一个是幻影3－4K——幻影3系列的一款新型无人机，能够进行超高清视频拍摄。另一个是Inspire 1 Pro黑色款，一款装配有相机的无人机。

去年11月，这家深圳无人机制造商发布了它的第一款农用无人机MG－1，主要用来喷洒农药。

大疆公司建立于2006年，现在占据世界消费无人机市场的70%左右。2014年销量达到400000架。据《福布斯》报道，该公司总收益达5亿美元，净利润1.2亿美元。大部分订单来自海外市场，2014年国内销量仅占10%。

大概一个月前，德国汉莎航空公司与大疆签订了一份协议，旨在开发用于检查飞机表面和监控风力发电场的商业无人机市场。

据北京一家咨询公司易观国际称，国内民用无人机市场预计将从今年的39.5亿元到2018年上升到110.9亿元。民用无人机分为消费层面的和工业层面的两种。业内人士分析，当前民用无人机的增长主要来自消费层面的无人机，主要用在娱乐行业，比如空中摄影。他们说工业层面的无人机会更有价值。

在消费电子产品展销会上，另一个中国无人机制造商亿航公司公布了世界上第一台能够携带一名人类乘客的无人机。该无人机充电时间仅两小时，能够以100km的时速在海平面以上飞行23分钟。公司希望这款无人机能够为乘客提供一种更加环保的出行方式，同时，帮助缓解交通堵塞。

随着无人机技术的进步以及政府的政策支持，将来无人机可能会用于矿产勘探、交通管理、灾情监测以及农业等领域。

Keys to Exercises of the Text

Ⅱ.
1. 无人机　　　　　　2. net profit
3. 高清视频拍摄　　　　4. civil drone
5. 矿产勘探　　　　　　6. disaster surveillance

Ⅲ.

drone; aircraft; Compared; dirty; originated; applications; surveillance; outnumber

Ⅳ.
1. 许多影院都配备了最先进的银幕和音响系统，而且周边的小吃也非常可口。
2. 小米无人机和大疆幻影3无人机的特征几乎是一样的，它们都有自动返回的功能，并且能够围绕某点运动。
3. 三星近日和国内的银联合作，正式发布了手机钱包服务。

名人传记阅读材料（Ⅱ）　　　　　　　　比尔·盖茨

威廉（比尔）H. 盖茨是微软公司主席、首席软件设计师，也是全世界个人及商务计算机领域软件制作、服务、互联网技术的领军人物。微软公司现有4万多名员工分布在世界上60多个国家。截止到2001年6月底本财政年度总收入为253亿美元。

盖茨1955年10月28日出生，与他的两个姐妹一起在西雅图长大。盖茨小学就读于一所公立学校，中学转到雷克赛德学校，这是一所私立中学。在那里，盖茨对软件发生了兴趣，13岁时便开始为计算机编制程序。

1973年，盖茨就读于哈佛大学，和史蒂夫·鲍尔默住在同一栋宿舍楼里，后者现在是微软公司的首席执行官。在哈佛读书期间，盖茨为第一台微机——MITS Altair编制了一套BASIC语言。

大学三年级时，盖茨从哈佛退学，专心于创建微软公司。早在1975年，盖茨便与童年时的朋友保罗·艾伦一起筹建微软公司。他们坚信计算机将是每张办公桌上和每个家庭里非常有用的工具。带着这一信念，他们着手开发个人计算机软件。盖茨对个人计算机的远见卓识是微软公司和软件业成功的关键。

1999年，盖茨写了一本名为《未来时速》的书。书中介绍了计算机技术怎样以全新的方式解决商务问题。此书有25种语言的版本，在60多个国家均有售。《未来时速》得到评论界的广泛赞誉，被《纽约时报》《今日美国》《华尔街日报》和"亚马逊"网站列入各自的畅销书目录。盖茨早些时候（1995）出版的另一本书《未来之路》曾连续七周高居《纽约时报》畅销书排行榜榜首。

除了对计算机和软件情有独钟外，盖茨对生物技术也有兴趣。他是数家生物技术公司的投资人,还创立了致力于开发世界上最大的视觉信息资源的考比司公司。此外,盖茨与移动电话的先驱克拉·麦考共同投资了 Teledesic 公司,这家公司正在实施一个雄心勃勃的计划:使用数百颗低轨道卫星为全世界的用户提供双向的宽带远程通信服务。

Keys to Exercises of the Reading

1. Inthe private Lakeside school he found his interest in software, and began programming computers at the age of 13.

2. Hedeveloped a version of programming language BASIC for the first microcomputer.

3. Withhis childhood friend Paul Allen, Gates started to build Microsoft.

4. Gates' anticipation and his vision for personal computing have been central to the success of Microsoft and the software industry.

5. Gates also founded Corbis, which is developing one of the world's largest resources of visual information.

Appendix 3　New Words List
（生词表）

a permanent load-side fault　永久负载侧故障
a revolving rotor　转子
a small air gap　气隙
a stationary stator　定子
abort　[əˈbɔt]　v. 中止；夭折
AC (Alternating Current)　交流
AC-bridge　交流电桥
accelerate　[əkˈseləreit]　vt. 使……加快，vi. 加速，促进，增加
accommodate　[əˈkɔmədeit]　vt. 供应，容纳，vi. 适应，调解
accompany　[əˈkʌmpəni]　vt. 陪伴，伴随，伴奏，vi. 伴唱
accomplished　[əˈkɔmpliʃt]　adj. 熟练的，完成的，有修养的，有学问的
accumulate　[əˈkjuːmjuleit]　vi. 累积，积聚，vt. 积攒
acquisition　[ˌækwiˈziʃən]　n. 获得，购置
actuator　[ˈæktjueitə]　n. 执行机构，激励者，促动器
adaptation　[ˈədæpˈteiʃən]　n. 适合，适应，适应性控制
adapter　[əˈdæptə]　n. 适配器，改编者
adhesive　[ədˈhiːsiv]　adj. 带粘性的，粘着，n. 粘合剂，胶黏剂
adjacent　[əˈdʒeisnt]　adj. 毗连的，接近的，相接触的
adjoin　[əˈdʒɔin]　vt. 贴近，毗连，毗邻
Administration sub-system　管理子系统
administrative　[ədˈministrətiv]　adj. 行政的，管理的
admittance　[ədˈmitəns]　n. 进入，入场权，通道，导纳（即电阻的倒数）
adopt　[əˈdɔpt]　vt. 收养，采用，采纳，采取，正式接受，通过
adoption　[əˈdɔpʃən]　n. 采用，收养
affirm　[əˈfəːm]　vt. & vi. 断言，证实
algebraic　[ˈældʒiˈbreiik]　adj. 代数的，关于代数学的
algebraic equation　差分方程
algebraic product　代数乘积
algebraical　[ˈældʒiˈbreiikəl]　adj. 代数学的（= algebraic）
algorithm　[ˈælgəˌriðəm]　n. 算法，运算法则
algorithm's recommendation　数字系统指令
alignment　[əˈlainmənt]　n. 定线，准线
alphanumeric　[ˈælfənjuːˈmerik]　adj. 字母数字的
alternating current (AC)　交流，交流电
alternator　[ˈɔːltəneitə]　n. 交流发电机
amplifier　[ˈæmplifaiə]　n. 放大器，扩大器，扩音器

amplify ['æmplifai] vt. 放大，扩大，增强，详述
amplitude ['æmplitju:d] n. 振幅，广阔，丰富，充足
amplitude modulation （AM）调幅
analog ['ænələɡ] n. 类似物，模拟 adj. 有长短针的，模拟的
analogous to 类似于……
analogue ['ænələɡ] n. 类似物，类似情况，adj. 类似的，相似物的
analogue controllers 模拟控制器
and gate 和门，与门，与电路
anti-clockwise adj. 逆时针方向的，逆时钟方向的
appreciable [ə'pri:ʃiəbl] adj. 可感知的，相当可观的，可评估的
apprehension ['æpri'henʃən] n. 理解，逮捕，恐惧，忧惧
arbitrary ['ɑ:bitrəri] adj. 任意的，武断的，n. 任意角度/形状
architecture ['ɑ:kitektʃə] n. 建筑学，建筑式样，建筑风格
armature ['ɑ:mətʃə] n. 电枢，转子，电枢线圈，衔铁线圈
arrester [ə'restə] n. 捕拿者，止动器，避雷装置
ascertain [æsə'tein] vt. 查明，弄清，确定
assembly [ə'sembli] n. 集合，装配，汇编，集会，组装
assembly line 流水线，装配线
assessment [ə'sesmənt] n. 确定，评定，评价，看法，核定的付款额
associated with 与……有关系，与……相联系
attenuation [ə,tenju'eiʃən] n. 衰减，变薄，稀释
attenuator [ə'tenjueitə] n. 衰减器，弱化子
audio ['ɔ:diəu] adj. 声音的，音频的，声频的
augment ['ɔ:gmənt] vi. & vt. n. 增加，增大
automatically ['ɔ:tə'mætikəli] adv. 自动地，机械地 adi. 不经思索的
auto-restoration algorithm 主动恢复数字系统装置
auto-sectionalizing switches 主动分段开关
auxiliary [ɔ:g'ziljəri] adj. 辅助的，补充的，备用的
backbone ['bækbəun] n. 决心，毅力，支柱，脊椎，主干网
backing film 底片
balance beam 平衡杆
battery-backed memory 随机存储器
be constituted with 由……组成
be coupled to 与……联结
be fastened to 被固定于…
be sensitive to 对……敏感
be subject to 遭受
be transferred to 转换
beam ['bi:m] n. 梁，（光线的）束，电波，v. 播送
bearing ['bεəriŋ] n. 承载，关系，方向，轴承，方位，举止，关系
beep [bi:p] vi. 嘟嘟响，n. 哔哔的声音，警笛声

binary ['bainəri] adj. 二元的，二态的，二进制的
binary arithmetic 二进制算术
bistable [bai'steibl] adj. 双隐定的，双稳态的
bottleneck ['bɔtlnek] n. 瓶颈，障碍物
BPS（Bits Per Second） 位/秒
breadboard ['bredbɔːd] n. 案板，电路试验板
breaker ['breikə] n. 断路器
building intercom system 楼宇对讲系统
bumplessly ['bʌmplisli] adv. 无扰动地
burglar ['bəːglə] n. 窃贼，破门盗窃者
burglar alarm system 防盗报警系统
buzzer ['bʌzə] n. 蜂鸣器，信号手
cabinet ['kæbinit] n. 橱，陈列柜，内阁，内阁会议
cable ['keibl] n. （船只、桥梁等上的）巨缆，钢索，电缆
cable television 有线电视，电缆电视
cabling ['keibliŋ] n. 卷缆柱，卷绳状雕饰
calibrate ['kælibreit] vt. 标定，分度，调整，校正
calibrated scale 刻度盘
capacitance [kə'pæsitəns] n. 电容，电流容量
capacitor [kə'pæsitə] n. 电容器
carbon ['kaːbən] n. 碳，碳棒，复写纸，adj. 碳的，碳处理的
carrier generation 载波
carry on 经营，从事，忙于，继续进行
cascade [kæ'skeid] n. 串联，级联
cascade control 串级控制
cast aluminum rotor 铸铝转子
catalyst ['kætəlist] n. 催化剂，刺激因素
category ['kætigəri] n. 种类，分类，范畴
cathode ['kæθəud] n. 阴极，负极
cathode-ray tube（CRT） 阴极射线显像管
central programming unit（CPU） 中央处理器
centrifugal [sen'trifjugəl] adj. 离心的，远中的，n. 转筒，离心机
centrifugal governor 离心式调速器
chip [tʃip] vt. 削，凿，vi. 碎裂，剥落，n. 碎片，芯片
chopper ['tʃɔpə] n. 斩波器
circuit breaker 断路器
circulate ['səːkjuleit] vt. （使）循环，（使）流通，（使）流传，散布，传播
circumference [sə'kʌmfərəns] n. 圆周，周长
closed-circuit television monitoring system 闭路电视监控系统
closed-loop control system 闭环控制系统
coat [kəut] vt. 为某物涂抹

coaxial cable　同轴电缆
coefficient　[ˌkəui'fiʃənt]　n. 协同因素，系数，adj. 合作的，共同作用的
coil　[kɔil]　v. 盘绕，把……卷成圈，n. 卷，线圈
combination　[ˌkɔmbi'neiʃən]　n. 结合，化合物，联合，组合
combustion　[kəm'bʌstʃən]　n. 燃烧
commit　[kə'mit]　vt. 保证
Communication Automation System　通信自动化系统
compatibility　[kəm'pæti'biliti]　n. 兼容性
compensate　['kɔmpenseit]　vi. 补偿，赔偿，抵消，vt. 付报酬
competence　['kɔmpitəns]　n. 能力，技能，(法院的)权限，管辖权
component　[kəm'pəunənt]　n. 成分，元件，组件，adj. 组成的，构成的
composed　[kəm'pəuzd]　adj. 镇静的，沉着的，v. 组成，著作
comprehensive　['kɔmpri'hensiv]　adj. 广泛的，综合的
computation　[kɔmpju:'teiʃn]　n. 计算，估计
condenser　[kən'densə]　n. 冷凝器，压缩器，聚合器，电容器
conductivity　[ˌkəndʌk'tiviti]　n. 传导性，导电性，传导率
configure　[kən'figə]　v. 配置，设定，使成形，使具一定形式
conform　[kən'fɔ:m]　vi. 遵守，符合，顺应，一致
conform to　符合，遵照
confusion　[kən'fju:ʒən]　n. 混乱，混淆
consent speed　额定转速
consequence　['kɔnsikwəns]　n. 结果，推理，推论，重要的地位
consequently　['kɔnsikwəntli]　adv. 从而，因此
conservation　[ˌkɔnsə'veiʃən]　n. 保存，保持，守恒
constant amplitude　等幅振荡
constitute　['kɔnstitju:t]　vt. 组成，构成，任命，建立
consumption　[kən'sʌmpʃən]　n. 消费，消耗
contact　['kɔntækt]　n. 接触器，电气接头，触头
contained　[kən'teind]　adj. 泰然自若的，从容的，v. 包含，容纳
contract agreement　合同协议书
control elements　控制元件
control valve　控制阀
controlled variable　被控量
Controller System for Industrial Automation　工业自动化控制器系统
conventional　[kən'venʃənl]　adj. 依照惯例的，约定俗成的
cooler　['ku:lə]　n. 冷却器，清凉剂
coordination　[kəu'ɔ:di'neiʃən]　n. 对等，同等，协调，调和
cord　[kɔ:d]　n. (细)绳，灯芯绒裤
core　[kɔ:]　n. 核心，果心，要点，磁心
correspondence　['kɔris'pɔndəns]　n. 信件，函件，通信，一致，相似
corresponding　['kɔris'pɔndiŋ]　adj. 相当的，对应的，符合的，一致的

| counteract | [ˌkauntəˈrækt] | vt. 抵消，中和，阻碍 |

counteract [ˌkauntəˈrækt] vt. 抵消，中和，阻碍
counterpart [ˈkauntəpɑːt] n. 副本；配对物；极相似的人或物
cram [kræm] vt. 塞入，填塞，塞满，（为考试而）死记硬背功课
crank [kræŋk] n. 曲柄
crime prevention 犯罪预防
criteria [kraiˈtiəriə] n. (单) 尺度，标准
cryogenic [ˌkraiəˈdʒenik] adj. 低温学的
crystal [ˈkristəl] n. 水晶，晶体，水晶饰品，adj. 水晶的，透明的
culmination [ˌkʌlmiˈneiʃən] n. 顶点，极点
current meter 电流表
cylindrical [siˈlindrikəl] adj. 圆柱形的，圆柱体的
damper [ˈdæmpə] adj. 变形 n. 令人沮丧（或扫兴）的人（或物），气流调节器
damping [ˈdæmpiŋ] n. 阻尼；衰减，减幅，抑制
data information station 数据信息点
DC (Direct Current) 直流
DC generator 直流发电机
decibel [ˈdesibel] n. 分贝
decimal constant 十进制常数
decreasing amplitude 衰减振荡
decrement [ˈdekrimənt] n. 减少量
defect [ˈdiːfekt, diˈfekt] n. 缺点，缺陷，vi. 叛变，变节
defective [diˈfektiv] adj. 有缺点的，不完美的，不完全的
definite [ˈdefinit] adj. 确切的，一定的，确定的，明确的
deflect [diˈflekt] vt. 使偏斜，使转向，使弯曲，vi. 偏斜，转向
deflection [diˈflekʃən] n. 偏向，偏差，挠曲
deliberation [diˌlibəˈreiʃən] n. 深思熟虑，研究，从容，沉着
denote [diˈnəut] vt. 表示，指示，意指
depict [diˈpikt] vt. 描画，描述，描写，叙述
deposit [diˈpɔzit] n. 堆积物，沉淀物，v. 存放，堆积，沉淀
derivative [diˈrivətiv] n. 衍生物，派生物，adj. 引出的，派生的
designate [ˈdezigneit] vt. 指明，称为，标志
detect [diˈtekt] vt. 发现，发觉，查明
detector [diˈtektə] n. 检测器，侦察器，发现者
deterioration [diˌtiəriəˈreiʃn] n. 恶化，降低，退化
deterrent [diˈterənt] n. 威慑力，制止物
deterrent measure 警戒措施
deviation [ˌdiːviˈeiʃən] n. 背离，偏离
DFRS (Digital Flight Recorder System) 数字故障记录仪
diagnose [ˈdaiəgnəuz] vt. 诊断，断定，vi. 判断，诊断
diagram [ˈdaiəgræm] n. 图表，图解，vt. 用图解法表示
diaphragm [ˈdaiəfræm] n. 横隔膜，隔膜，隔板，快门，光圈

dielectric　［ˌdaiiˈlektrik］　*adj.* 非传导性的，*n.* 电介质，绝缘体
dielectric constant　介电常数，电容率
diesel　［ˈdiːzəl］　*n.* 柴油机，柴油
diesel and jet engine　柴油机和喷气发动机
digital　［ˈdidʒitəl］　*adj.* 数字的，手指的，*n.* 数字，键
digital buses　数字总线
digital fault recorder　数字故障记录仪
digital instrument　数字仪器
digital simulation　数字信号仿真
dimension　［diˈmenʃən, dai-］　*n.* 维，尺寸，*vt.* 标出尺寸
discipline　［ˈdisiplin］　*n.* 纪律，学科，*vt.* 训练
disconnect　［ˌdiskəˈnekt］　*vt.* 拆开，使分离，*vi.* 断开
discrete　［disˈkriːt］　*adj.* 离散的，不连续的，*n.* 分立元件，独立部件
dispatcher　［disˈpætʃə］　*n.* 调度员，发报机
display　［ˌdisˈplei］　*n.* 显示（器），*vt.* 陈列，*adj.* 展览的
dispose　［disˈpəuz］　*vt. & vi.* 处理，处置，布置
disposition　［ˈdispəˈziʃən］　*n.* 气质，天性，性格，安排，布置
dissipation　［ˌdisiˈpeiʃən］　*n.* 消耗，分散
dissipative　［ˈdisipeitiv］　*adj.* 消耗的，浪费的，消散的
distinction　［disˈtiŋkʃən］　*n.* 差别，区别，特性，荣誉
distribute　［disˈtribju(ː)t］　*vt.* 散布，分布
Distributed Control Systems　分布式控制系统
distribution　［distriˈbjuːʃən］　*n.* 分配，分布
distribution box　配线箱
division　［diˈviʒən］　*n.* 分，分割，划分，分化现象
document flow　文档流程
domain　［dəuˈmein］　*n.* 领域，产业，地产，域名
doped-semiconductor　掺杂半导体
dot　［dɔt］　*n.* 点，圆点，嫁妆，*vi.* 打上点，*vt.* 加小点于
drop-out current　释放电
dual　［ˈdjuːəl］　*adj.* 两部分的，二体的，二重的
dual-trace oscilloscope　双踪示波器
duplicate　［ˈdjuːplikeit］　*vt.* 复制，复印，*adj.* 完全一样的
duress　［djuəˈres］　*n.* 威胁，逼迫
dynamics　［daiˈnæmiks］　*n.* 动力学，力学
eddy　［ˈedi］　*n.* 漩涡，涡流，逆流，*vi.* 旋转，起漩涡，*vt.* 使……起漩涡
eddy current　涡轮电流
EEPROM　电可擦写可编程只读存储器
elapse　［iˈlæps］　*vi.* 经过
Electrical Rule Check　电气规则检查
electromagnetic　［iˈlektrəumægˈnetik］　*adj.* 电磁的

electronic counter　电子计数器
electronic design application　电子设计软件
electronic transformer　电力变压器
Electronics Workbench　电子工作台
electrostatic　[iˌlektrəˈstætik]　adj. 静电学的，静电的
element　[ˈelimənt]　n. 元素，成分，要素，原理，自然环境
emf　abbr. n. 电动势（Electromotive Force）
emulation　[ˌemjuˈleiʃən]　n. 竞争，仿真
encounter　[inˈkauntə]　vt. 遭遇，邂逅，遇到　n. & vi. 偶然碰见
enhancement　[inˈhɑːnsmənt]　n. 增强，提高，放大
enterprise　[ˈentəpraiz]　n. 事业心，进取心，企［事］业单位
equivalent　[iˈkwivələnt]　adj. 等价的，相等的，同意义的　n. 等价物
erosion　[iˈrəuʒən]　n. 腐蚀，侵蚀，磨损
error detection element　误差检测元件
estate　[isˈteit]　n. 土地，地区，庄园，种植园，地产，财产，遗产
etch　[etʃ]　vt. & vi. 用针和酸类在金属板上蚀刻（图画等）
Ethernet Switch　以太网交换机
evaporate　[iˈvæpəreit]　v. 蒸发，挥发，消失，使蒸发，使挥发
excitation　[eksiˈteiʃn]　n. 激发，励磁，刺激
excitation current　励磁电流
exciting voltage　励磁电压
execution　[ˌeksiˈkjuːʃən]　n. 执行，实行，完成，死刑
existing technology　现有技术
expressway　[ikˈspreswei]　n. 高速公路
facilitate　[fəˈsiliteit]　vt. 促进，帮助，使容易
facility　[fəˈsiliti]　n. 设备，便利的设施
facsimile　[fækˈsimili]　n.（文字、图画等的）副本，传真
faculty　[ˈfækəlti]　n. 能力，才能，院，系，部，全体从业人员
fall short　未能满足，不能达到
farad　[ˈfærəd]　n. 法拉（电容单位）
fault current　故障电流
feedback　[ˈfiːdbæk]　n. 反馈，反应
feedback control systems　反馈控制系统
file management　档案管理
film　[film]　n. 胶卷，薄膜，轻烟，膜层，vi. 生薄膜
filter　[ˈfiltə]　vi. 慢慢传开，滤过，n. 滤波器，筛选，vt. 过滤，渗透
filtering　[ˈfiltəriŋ]　v. 过滤，滤除（filter 的 ing 形式）
fin　[fin]　n. 散热片
final control element.　末控制元件
fire alarm system　消防报警系统
flip-flop　[ˈflipflɔp]　n. 触发器，啪嗒啪嗒的响声，vt. 使翻转

fluorescent　[fluəˈresnt]　adj. 荧光的，容光焕发的，n. 荧光，荧光灯
flux　[flʌks]　n. 磁力线，磁通量
focus　[ˈfəukəs]　n. 焦点，焦距，中心，v.（使）聚焦，n. 焦点，焦距
foil　[fɔil]　n. 箔，金属薄片，叶形片，烘托，vt. 衬托，阻止，贴箔于
form wound　模绕
fossil-fuel plant　燃料电厂
frequency meter　频率计
frequency modulation　（FM）调频
frequency response　频率响应
frequency synthesizer　频率合成器
full supervisory control　全自动监控
function　[ˈfʌŋkʃən]　n. 功能，函数，盛大的集会，vi. 行使职责，运行
functional　[ˈfʌŋkʃənl]　adj. 功能的，职能的，函数的
fungus　[ˈfʌŋgəs]　n. 真菌
furnace　[ˈfəːnis]　n. 熔炉，火炉
fuse　[fjuːz]　n. 熔丝
gain　[gein]　n. 收获，增益，利润，vt. 获得，赚到，vi. 获利，增加
garage　[ˈgærɑːdʒ]　n. 车库，汽车修理厂，飞机库，vt. 把……送入车库
gauge factor　量规因数
generator　[ˈdʒenəreitə]　n. 发电机，发生器
generator action　发电机作用
geometric　[dʒiəˈmetrik]　adj. 几何学图形的，几何学的
give rise to　导致
glowing　[ˈgləuiŋ]　adj. 灼热的，v. 发光，容光焕发
graded time settings　阶梯形时间配制
graphical waveform　波形分析
grid　[grid]　n. 格子，格栏，网格，方格
guideline　[ˈgaidlain]　n. 指导方针，准则
habitation　[ˌhæbiˈteiʃən]　n. 居住，住宅，家
hand crank　手摇曲柄
handcrafting-type method　手工方式
hazardous　[ˈhæzədəs]　adj. 冒险的，碰运气的，有危险的
hence　[hens]　adv. 因此，今后
henry　[ˈhenri]　n. 亨利　[电感单位，略作 H]
hood　[hud]　n. 头巾，兜帽，覆盖，vt. 以头巾覆盖，罩上
horizontal　[hɔriˈzɔntəl]　adj. 水平的，地平线的
horizontal sweep　水平扫描
horizontally　[hɔːriˈzɔntli]　adv. 水平地
horseshoe magnet　[ˈmægnit]　马蹄形磁铁
hot rail　火线
housing　[ˈhauziŋ]　n.（机器等的）防护外壳或外罩

human intervention　人类介入
HVAC　暖通空调系统
hydraulic　[haiˈdrɔːlik]　adj. 水力的，液压的，水力学的
hydroturbine　[ˌhaidrəuˈtəːbin]　n. 水轮机
hysteresis　[histəˈriːsis]　n. 滞后（现象），滞后作用
idealize　[aiˈdiəlaiz]　vt. 使理想化
IEC　国际电工委员会
illumination　[iˌljuːmiˈneiʃən]　n. 照明，强度，彩灯，灯饰
illustrate　[ˈiləstreit]　vt. 阐明，表明，显示为例证
impedance　[imˈpiːdəns]　n. 全电阻，阻抗
implement　[ˈimplimənt]　vt. 使生效，贯彻，执行，n. 工具，器具，用具
implementation　[ˌimplimenˈteiʃən]　n. 履行，实现，安装启用
import and export control system　进出口控制系统
impulse　[ˈimpʌls]　n. 冲动，脉冲，刺激，神经冲动，vt. 推动
in accordance with　依照，根据
In brief　简言之，简单地说
in line　adv. 成一直线，有秩序，协调，adj. 联机的
in parallel with　与……并联
in phase with　与……相同
in series with　与……串联
in the event　结果，如果
incorporation　[inˌkɔːpəˈreiʃən]　n. 合并，结合
increasing amplitude　发散振荡
independent of　不受……支配的，与……无关的
induce　[inˈdjuːs]　v. 劝诱，促使，感应
induced　[inˈdjuːst]　adj. 感应的，引诱的
induced current　[inˈdjuːst]　感生电流
inductance　[inˈdʌktəns]　n. 感应系数，自感应，电感
induction-disc relay　感应圆盘式继电器
inductor　[inˈdʌktə]　n. 授职者，感应器
inferential measurements　推理性测量
inferior　[inˈfiəriə]　adj. 低等的，下级的，劣等的，次的，n. 部下，下属
information outlet　信息插座
infrastructure　[ˈinfrəˈstrʌktʃə]　n. 基础设施，基础结构
infrastructure　[ˈinfrəˌstrʌktʃə]　n. 基础构造，基础结构
inherent　[inˈhiərənt]　adj. 固有的，内在的，与生俱来的，遗传的
initialize　[iˈniʃəlaiz]　vt. 初始化
initiate　[iˈniʃieit]　v. 创立，引进
innovative　[ˈinəuveitiv]　adj. 新发明的，新引进的，革新的
insert　[inˈsəːt, ˈinsəːt]　vt. 插入，嵌入，n. 插入物
installation　[ˈinstəˈleiʃən]　n. 安装，装置，设备，军事设施

installed [in'stɔ:ld] adj. 安装的，已装入的（install 的过去分词）
instantaneous ['instən'teinjəs] adj. 瞬间发生的，即刻的
institution ['insti'tju:ʃən] n. 惯例，习俗，制度，慈善机构
instrumentation [ˌinstrumen'teiʃn] n. 仪表化，测试设备
insulate ['insjuleit] vt. 隔离，使孤立，使绝缘，使隔热
insulating adj. 绝缘的，隔热的
insulation [insju'leiʃən, insə'leiʃən] n. 隔离，孤立，绝缘
integrate ['intigreit] vt. 使……成整体，vi. 求积分，成为一体，adj. 整合的
integrated services digital network 综合业务数字网
integrity [in'tegrəti] n. 正直，诚实，廉正，完整
intelligent [in'telidʒənt] adj. 聪明的，理解力强的，智能的
intelligent home system 智能家居系统
intelligent system 智能系统
intensity [in'tensəti] n. 强烈，强度，亮度，紧张
interconnect [ˌintə(:)kə'nekt] vt. 使互相连接，vi. 互相联系
interconnection [ˌintəkə'nekʃən] n. 互相联络，互联
interface ['intəfeis] n. 接口，界面
internal battery 内装电池
interruption [intə'rʌpʃən] n. 停止，中断
intruder [in'tru:də] n. 未请自入者，闯入者（尤指企图行窃者）
inventory ['invəntəri] n. 详细目录，存货清单
inversely ['invə:sli] adv. 倒转地，相反地，反比例地
inversion [in'və:ʃən] n. 倒置，倒转，反向反转
invert [in'və:t] a. 转化的，v. 反转，颠倒，反置
inverting input 反向输入端
involved [in'vɔlvd] adj. 卷入的，有关的（involve v. 涉及，包含）
irreplaceable [iri'pleisəbl] adj. 不可替代的，独一无二的
jack [dʒæk] n. 插孔，插座，起重器，vt.（用起重器）抬起
judicious [dʒu:'diʃəs] adj. 有见识的，明智的
Kelvin n. 绝对温标，开氏（开尔文）温标
knob [nɔb] n. 旋钮
ladder logic 梯形图
lag network 滞后网络
lead network 超前网络
laptop ['læptɔp] n. 膝上型轻便计算机
lay out 展开，设计，布置，划定
layout ['leiaut] n. 规划，编排，布局
lead resistance 引线电阻
leakage ['li:kidʒ] n. 漏，漏损物，泄漏，漏损量
lightning protectors 避雷保护
limit switch 行程开关

line to line fault　时间故障
link with　将……与……连接〔系〕在一起
lock-in amplifier　锁定放大器
logic gates　逻辑闸，逻辑门
loop　[lu:p]　vi. 打环，翻筋斗，n. 环，圈，（闭合）回路
lube　[lu:b]　n. 润滑油（等于 lubrication）
luxury　['lʌkʃəri]　n. 奢侈，豪华，奢侈品
magnet　['mægnit]　n. 磁铁，磁石，磁体
magnetic　[mæg'netik]　adj. 有磁性的，地磁的
magnetic field　磁场
magnetic stripe　（卡片或文件上的）磁条
magnetize　['mægnitaiz]　vt. 使磁化，吸引
maintenance　['meintinəns]　n. 维持，维护，保养，维修，赡养费
make certain　处理，应付
manipulate　[mə'nipjuleit]　vt. 操作，操纵，巧妙地处理，篡改
manipulation　[mə'nipju'leiʃən]　n. 操作，操纵，处理，篡改
manipulated variable　控制量
manufacture　[ˌmænju'fæktʃə]　n. 制造，制造业，v. 加工
manufactured good　制造商品
manufacturing　[mænju'fæktʃəriŋ]　adj. 制造业的，n. 制造业，v. 制造
mass production　批量生产
MDS（Microwave Data System）微波数据系统
mechanical andelectrostatic precipitators　机械和静电沉淀器
mechanism　['mekənizəm]　n. 机械装置，机制，进程
memory space　存储空间
merge　['mə:dʒ]　vt. & vi. （使）混合，（使）合并
mesh　[meʃ]　n. 网络，网眼，网丝，圈套，vi. 相啮合，vt. 以网捕捉
metallic　[mi'tælik, me-]　adj. 金属的，含金属的，n. 金属纤维
meter　['mi:tə]　n. 公尺，仪表，米，vt. 用仪表测量，vi. 用表计量
microfarad　[ˌmaikrəu'færəd]　n. 微法拉（电容量的实用单位）
minimize　['minimaiz]　v. 使减（缩）小到最低，最低估计
mitigate　['mitiˌgeit]　vt. 使减轻，使缓和
mnemonic　[ni:'mɔnik]　adj. 记忆的，助记的，记忆术的
modular　['mɔdjulə]　adj. 模数的，模块化的
modulate　['mɔdjuleit]　vt. 调节，调整，（信号）调制，vi. 转调
module　['mɔdju:l]　n. 模数，模块，组件
monitor　['mɔnitə]　n. 监视器，监听器，监控器，vt. 监控
monitoring　['mɔnitəriŋ]　n. 监视，控制，监测，追踪
multi-meter　万用电表，多量程仪表
multimeter　[mʌl'timitə]　n. 万用表，数字万用表，多用电表
multiple　['mʌltipl]　adj. 多样的，许多的，多重的，n. 并联，倍数

multiply	['mʌltiplai]	v. 乘，(使)相乘，adv. 多样地，adj. 多层的
multi-processor computer		多处理器计算机
multivibrator	[ˌmʌltivai'breitə]	n. 多谐振荡器
mutually	['mjuːtʃuəli]	adv. 相互地，彼此地
mutually interacting coils		相互作用线圈
nature	['neitʃə]	n. 自然，性质，种类，本性
needle	['niːdl]	n. 针，指针，针状物，刺激，vi. 缝纫，做针线
network office workbench		网络工作平台
neural	['njuərəl]	adj. 神经的
neutral rail		零线
node	[nəud]	n. 节点，瘤，结点
nodal		adj. 节的，波节的，结点的
non-faulted feeder zones		无故障供电范围
non-inverting input		同向输入端
nonlinearity	['nɔnlini'ærəti]	n. 非线性，非线性特征
non-paper office		无纸化办公
nonplanar		adj. 非平面的，空间的（曲线的）
notch	[nɔtʃ]	n. (表面上的) V 型痕迹，刻痕，vt. 在（某物）上刻 V 形痕
notify	['nəutifai]	vt. 通告，通知，公布
obsolete	['ɔbsəliːt]	adj. 不再使用的，过时的
obsolete armless insulator		过期无防护的绝缘器
occasion	[ə'keiʒən]	v. 致使，惹起，引起
ohmmeter	['əumˌmiːtə]	n. 电阻表，欧姆计
omega	['əumigə]	n. 希腊字母的最后一个字，终了，最后
on account of		因为，考虑
on file		存档
on the negative side		从消极方面看
on the positive side		从积极方面看
operational	[ˌɔpə'reiʃənəl]	adj. 操作的，运作的经营的
operational routines		操作程序
opposition	[ˌɔpə'ziʃən]	n. 反对，敌对，相反，反对派
optimize	['ɔptimaiz]	vt. 使最优化，使完善，优化，持乐观态度
optimum	['ɔptiməm]	adj. 最适宜的，n. 最佳效果，最适宜条件
organ	['ɔːgən]	n. 风琴，器官，机构，管风琴，嗓音
organic	[ɔː'gænik]	adj. 有机的，器官的，官能上的
orientation	[ɔːrien'teiʃən]	n. 方向，方位，定位
oscillate	['ɔsileit]	vt. 使动摇，使振动，使振荡，vi. 振荡，犹豫，摆动
oscillation	[ɔsi'leiʃn]	n. 摆动，震动
oscillator	['ɔsileitə]	n. 动摇不定的人，犹豫的人，振荡器
oscilloscope	[ɔ'siləskəup]	n. 示波器
outage	['autidʒ]	n. 停机，断电

overdue ['əuvə'djuː] adj. 迟到的，延误的，过期的，到期未付的
overdue documents　过期文档
PABX　专用自动交换分机
package ['pækidʒ] n. 程序包管壳，套装软件，vt. 将……包装，打包
panel ['pænəl] n. 面，板，控制板，仪表盘
parallel input / output　并行输入/输出
parameter [pə'ræmitə] n. 因素，特性，参量，参数
parameters [pə'ræmitə] n. 参数，参量，系数
partition [paː'tiʃən] n. 分割，划分，分开，隔墙，vt. 区分，分隔
passive components　无源器件
patch cord　跳线
patching case　配线箱
patent ['peitənt] adj. 有专利的，受专利权保护的
payroll ['peirəul] n. （员工的）工资名单，（公司的）工资总支出，工薪总额
peak-to-peak voltage　电压峰-峰值
perimeter [pə'rimitə] n. 周边，周围，边缘
perimeter guard system　周界防范系统
period ['piəriəd] n. 周期，期间，课时，adj. 某一时代的
permanent ['pəːmənənt] adj. 永久的，永恒的，不变的
perpendicular [ˌpəːpən'dikjulə] adj. 垂直的，正交的，直立的，n. 垂线
phase shift　相位漂移/差别，移相
phasor ['feizə(r)] n. 相量．相量图
philosophy [fi'lɔsəfi] n. 原理，哲学，哲理，人生观
phosphor ['fɔsfə] n. 磷光体，磷光剂
pickup Current　始动电流
picofarad ['piːkəu'færəd] n. 微微法拉（百亿分之一法拉），皮可法拉
piezoresistive effect　压阻效应
pin [pin] n. 大头针，钉，管脚，琐碎物，vt. 钉住，将……用针别住
8 pin dual-in line　8管脚双列直插式
platinum ['plætinəm] n. 白金，铂
plug [plʌg] n. 栓，插头，塞子，vi. 用插头将与电源接通，vt. 插入，接插头
plunger ['plʌndʒə] 可动铁心，活塞
pneumatic [njuː'mætik] adj. 气动的，有气胎的，充气的，n. 气胎
pointer ['pɔintə] n. 指针，暗示，指示器
polarity [pəu'lærəti] n. 两极，极性，对立
polarity switch　极性开关
potential [pəu'tenʃəl] n. 电势
power supply　电源
power system fault　电力系统故障
precaution [pri'kɔːʃən] n. 预防措施
precision [pri'siʃən] n. 精确，精密度，adj. 精密的，精确的

preload ['priː'ləud] vt. 预加载,预装入
premise [pri'maiz] n. 处所,楼宇,前提
preset [priː'set] adj. 预先装置的,预先调整的
prevai [pri'veil] vi. 流行,经常发生
primarily ['praimərəli] adv. 主要地,首先,起初
prime mover 原动机
primitive ['primitiv] adj. 原始的,早期的,简单的,粗糙的
principle ['prinsəpl] n. 原理,原则,道义,本质,本义,根源,源泉
prize [praiz] vt. 撬
probe [prəub] n. 探针,调查,vi. 调查,探测,vt. 探查,用探针探测
process automation and control 过程自动化和过程控制
process control 过程控制
program/data memory 程序/数据存储器
prohibitively [prəu'hibitivli] adv. 禁止地,过高地,过分地
proportional [prəu'pɔːʃənəl] adj. 成比例的,n. 比例项
proprietary [prəu'praiətəri] n. 所有权,所有人,adj. 所有的,专利的
protocol ['prəutəkɔl] n. 协议
providing [prə'vaidiŋ] conj. 假如,以……为条件
provision [prə'viʒən] n. 供应,提供,供给,规定,条款,条件
proximity [prɔk'simiti] n. 接近,附近
public address system 公共广播系统
pulse [pʌls] n. 脉搏,脉冲 v. 使跳动,脉跳
pulverizer ['pʌlvəraizə] n. 粉碎机,喷雾器,粉碎者
quadrature ['kwɔdrətʃə] n. 求积,矩,上(下)弦
radial ['reidiəl] adj. 放射状的,辐射状的,星形的
ramp generator 斜坡发生器
random ['rændəm] adj. 任意的,随机的,n. 随意
random access memory(RAM) 随机存储器
random wound 散绕
rate valve 速率阀
rating ['reitiŋ] n. (船上人员的)等级,类别,(海军)水兵
ratio ['reiʃiəu] n. 比率,比例
read only memory(ROM) 只读存储器
real estate 房地产,房地产所有权
real-time parameter 实时参数
reciprocal [ri'siprəkəl] adj. 相互的,倒数的,n. 倒数
rectangular ['rek'tæŋgjulə] adj. 矩形的,成直角的
rectifier ['rektifaiə] n. 整流器,改正者,矫正者
reduction [ri'dʌkʃən] n. 减少,缩减量
refer to as 提到……,作为……,把……称为
relaxation oscillator 弛缓振荡器

relay [ri'lei]   v. 转播，接替，n. 继电器，接替，接替人员
release current   释放电流
reliability [ri͵laiə'biliti]   n. 可靠性
remove [ri'mu:v]   vt. 移动，开除，调动，vi. 搬家，n. 距离
renovation ['renə'veiʃən]   n. 翻新，修复，整修
repetitive [ri'petətiv]   adj. 重复，迭代
represent [͵repri'zent]   vt. vi. 表现，描绘，代表，再赠送
reproducible [ri:prə'dju:səbl]   adj. 可再生的，可繁殖的，可复写的
residential [͵rezi'denʃəl]   adj. 提供住宿的，居住的，住宅的
resin ['rezin]   n. 树脂，松香 vt. 涂树脂，用树脂处理
resistive [ri'zistiv]   adj. 有抵抗力的，抵抗的，n. 电阻式
resistivity [͵ri:zis'tiviti]   n. 电阻系数，电阻率
resolution [͵rezə'lu:ʃən]   n. 分辨率
respect [ri'spekt]   n. 尊敬，方面，vt. 尊敬，尊重，遵守
responsivity [ri'spɔnsiviti]   n. 响应度，敏感度，响应率
restoration [restə'reiʃn]   n. 回复到原处或原状，恢复，修复，整修
restructure [ri:'strʌktʃə]   v. 重新组织，调整
result in   结果是，导致
retain [ri'tein]   vt. 保持，雇，记住
revolving horseshoe magnet   旋转的马蹄形磁铁
rms voltage   电压有效值
robustness [rəu'bʌstnis]   n. 健壮性，稳健性，鲁棒性
rotate [rəu'teit]   v. 旋转，转动
rotating magnetic field   旋转磁场
rotational [rəu'teiʃənəl]   adj. 转动的，回转的，轮流的
rotor ['rəutə]   n. 转子，回转轴，转动体
RTU (a Remote Terminal Unit)   远端电源设备
SCADA (Supervisory Control And Data Acquisition)   监控数据探测
scale [skeil]   n. 刻度，衡量，数值范围，v. 依比例决定，攀登
schematic [ski:'mætik]   adj. 图解的，概要的，n. 图解视图，原理图
schematic editor   电路编辑软件
schematic [ski:'mætik]   adj. 扼要的，图解的
screwdriver [͵skru:'draivə]   n. 螺钉旋具
semiconductor [͵semikən'dʌktə]   n. 半导体
sensor ['sensə]   n. 传感器
sequential [si'kwenʃəl]   adj. 连续的，相继的，有顺序的
Sequential Function Charts   顺序功能图
serial input/output   串行输入/输出
serially ['siəriəli]   adv. 逐次地，连载地，串行地
series trip coils   串联跳闸线圈
serpentine ['sə:pəntain]   adj. 蜿蜒的，弯弯曲曲的

service interruption 停电
servomechanism [ˌsəːvəuˈmekənizəm] n. 伺服机构，自动控制装置
setup [ˈsetʌp] n. 设置，装备，组织，计划，机构，调整
shaft [ʃæft] n. 连杆，传动轴，旋转轴
sharp [ʃɑːp] adj. 急剧的，锋利的，强烈的，adv. 锐利地，n. 内行，尖头
shed [ʃed] v. 去掉，除掉，脱落，剥落，蜕下
short-circuiting ring 短路环
shuttle power 往复能量
shuttle [ʃʌtl] n. 梭形，往复
signal modifier 信号调节器
signal-conditioning 信号调节，信号波形加工（修整）
simulation [ˌsimjuˈleiʃən] n. 模拟，仿真，模仿，假装
simulator [ˈsimjuleitə] n. 仿真器
simultaneous [ˌsiməlˈteiniəs] adj. 同时的，同时发生的，n. 同时译员
single-stage [ˈsiŋglˈsteidʒ] adj. 单级，单级的
sinusoid [ˈsainəsɔid] n. 正弦曲线（sinusoidal adj. 正弦曲线的）
sketch [sketʃ] n. 素描，略图，梗概，v. 画素描或速写
sliding contact 滑动触点
slip ring 滑环
slot [slɔt] n. 狭缝，狭槽，vt. 把……放入狭长开口中，把……纳入其中
slow-speed overcurrent relays 延时过电流继电保护
smart parking management system 智能停车场管理系统
solely [ˈsəuli] adv. 独自地，单独地
solenoid [ˈsəulənɔid] n. 螺线管，螺线形电导管
solenoid relay 螺管式继电器
sophisticated [səˈfistikeitid] adj. 老练的，老于世故的，精密的，尖端的
specific [spiˈsifik] adj. 特定的，明确的，详细的，具有特效的
specification [ˌspesifiˈkeiʃən] n. 规格，详述，说明书
spectrum [ˈspektrəm] n. 光谱，频谱，范围，幅度
spectrum radio network 波谱无线电网络
sphere [sfiə] n. 范围，球体，vt. 放入球内，使……成球形，adj. 球体的
spike [spaik] n. 长钉，尖峰信号，vt. 用尖物刺穿
spring [spriŋ] n. 弹簧，发条
stability [stəˈbiliti] n. 稳定性
stabilize [ˈsteibilaiz] vt. 使稳固，使安定，vi. 稳定，安定
state-of-the-art 尖端科技
stator [ˈsteitə] n. 定子，固定片
status [ˈsteitəs] n. 情形，状况，身份，地位
steam and hydroturbine 汽（水）轮机
steam-generating unit 蒸汽发生器
steer [stiə] vt. 驾驶，控制，引导，vi. 驾驶，掌舵，行驶

step down   逐渐缩小，降低
step up   增加某事物，促进某事物
step-up transformer   升压变压器
stereo   ['stiəriəu]   n. 立体声音响器材，立体声
stipulate   ['stipjuleit]   vi. 规定，保证，vt. 规定，adj. 有托叶的
strain gauge   应变仪，变形测量器
stray capacitance   寄生电容，杂散电容
stripe   [straip]   n. 条纹，斑纹，种类，vt. 加条纹于……
Structured Cabling System   综合布线系统
structured module   布线模块
substation   ['sʌbsteiʃən]   n. 变电站
substation breaker status   变电站断路器状况
substrate   ['sʌbstreit]   n. 基质，基片，衬底（等于 substratum）
subtle   ['sʌtl]   adj. 微妙的，细微的，狡猾的，狡诈的，敏感的，敏锐的
subtract   [səb'trækt]   vt. 减去
subtraction   [səb'trækʃən]   n. 减少，减法，差集
superimposed   ['sju:pəim'pəuzd]   adj. 上叠的，重叠的，叠加的
supervise   ['sju:pəvaiz]   v. 监督，管理，指导
supreme   [sju:'pri:m]   adj. 最高的，至上的，最重要的
susceptible   [sə'septəbl]   adj. 易受影响的，易感动的，n. 易得病的人
sustain   [sə'stein]   vt. 支撑，承担，维持，忍受，供养，证实
sustained   ['sʌsteind]   adj. 持久的，经久不衰的
symmetrical position   平衡位置
synchrocyclotron   [ˌsinkrəu'saiklətrɔn]   n. 同步回旋加速器
synchronization   [ˌsiŋkrənai'zeiʃən]   n. 同一时刻，同步
synchronous   ['siŋkrənəs]   adj. 同步的，同时的
synonymous   [si'nɔniməs]   adj. 同义的，同义词的，同义突变的
synthesizer   ['sinθisaizə]   n. 合成者，合成器，合成仪，综合器
System Integration Center   系统集成中心
tank   [tæŋk]   n.（盛液体，气体的大容器）桶、箱、罐、槽
tap   [tæp]   n. 抽头，水龙头，轻打
telex   ['teleks]   n. 用户电报，电传系统，电传，vt. 以电传发出（消息）
test leads   表笔
the adjacent tie circuit   临近线路
the closed-switch state   开关闭合状态
the open-switch state   开关断开状态
the theory of process control   过程控制理论
theorem   ['θiərəm]   n. 定理，法则，定律
thermoelectric   ['θə:məui'lektrik]   adj. 热电的
thin-film   ['θinkfilm]   n. 薄膜
three-phase alternating current   三相交流电

three-phase windings 三相绕组
threshold ['θreʃhəuld] n. 极限，门槛，入口，开始，临界值
threshold voltage 阈值电压
time diagram 时序图
time-current characteristic 时间-电流特性
time-delay compensator 延时补偿
timer/counter facilities 定时器/计数器
timing components 时钟组件
to reach the desired level of service continuity 达到理想的持续供电标准
tone [təun] n. 音调，语气，色调，vt. 用某种调子说，vi. 颜色调和
topology [tə'pɔlədʒi] n. 拓扑，布局，拓扑学
torque [tɔ:k] n. 扭转力，转力矩
torque requirement 转矩要求
trace [treis] vi. 追溯，沿路走，vt. 追踪，n. 痕迹，踪迹
trace out 描绘出
trait ['treit] n. 人的个性，显著的特点，特征
transducer [trænz'dju:sə] n. 传感器，变换器，换能器
transfer [træns'fə:] n. 转移，转让，传递，v. 转移，转让
transformer [træns'fɔ:mə] n. 变压器
transient ['trænziənt] adj. 短暂的，路过的，n. 瞬变现象，候鸟
transient analysis 暂态分析
transition [træn'ziʃən] n. 过渡，转变，变迁
transition cost 交易成本
transmission [trænz'miʃən] n. 传输，播送，变速器，传递的信息
transmission and distribution system 输配电系统
transmission equipment 输电设备
transmission line 馈电线，输电线
transmitter [trænz'mitə] n. 发射机，发报机，传达人，传导物
traverse ['trævəs] n. 横木，穿过，vt. 穿过，详细研究，vi. 横越，旋转
troubleshoot ['trʌblʃu:t] vt. 检修故障，查找故障
troubleshooter ['trʌbl,ʃu:tə] n. 纷争解决者，调解纷争专家
truth table 真值表
tuning ['tju:niŋ] n. 调谐，调音，起弦，协调一致，起音，定音
turbine ['tə:bain] n. 涡轮机，汽轮机
twist [twist] vt. & vi. 扭，搓，缠绕，vt. 转动，拧，歪曲，曲解
ubiquitous [ju:'bikwitəs] adj. 普遍存在的，无所不在的
underlie [,ʌndə'lai] vt. 位于……之下，成为……的基础
unify ['ju:nifai] vt. 使联合，统一，使相同，使一致
valve [vælv] n. 阀，活门，vt. 装阀于，以活门调节
variable ['vɛəriəbl] adj. 易变的，多变的，n. 可变物，可变因素
variable ['vɛəriəbl] adj. 易变的，多变的，n. 可变物，变量

variable frequency 变频
variable resistor 可变电阻器
variable speed drive 变速器
variety [vəˈraiəti] n. 种类，多样，杂要变化，多种
versatile [ˈvəːsətail] adj. 多用途的，多功能的
versus [ˈvəːsəs] prep. 对，对抗
vertical [ˈvəːtikəl] adj. 垂直的，直立的，n. 垂直线，垂直面
virtually [ˈvəːtʃuəli] adv. 事实上，几乎，实质上
volatile [ˈvɔlətail] adj. 易变的
voltage divider 分压器
voltmeter [ˈvəultˌmiːtə] n. 伏特计，电压表
wafer [ˈweifə] n. 圆片，晶片，薄片，干胶片，薄饼，vt. 用干胶片封
water hammer 水锤
water tank lever control system 水箱水位控制系统
waveshaping circuits 整形电路
winding [ˈwaindiŋ] n. 绕组，线圈
wire [ˈwaiə] n. 金属丝，电线，vt. 拍电报
work area sub-system 工作区子系统

参 考 文 献

[1] 刘小芹,等. 电子与通信技术专业英语 [M]. 4版. 北京:人民邮电出版社,2014.
[2] 周伯清. 电子信息专业英语 [M]. 北京:科学出版社,2009.
[3] 朱一纶. 电子技术专业英语 [M]. 4版. 北京:电子工业出版社,2015.
[4] 江华圣. 电工电子专业英语 [M]. 2版. 北京:人民邮电出版社,2010.
[5] 林涌,刘宇红. 机电专业英语 [M]. 北京:化学工业出版社,2014.
[6] 徐存善. 自动化专业英语 [M]. 北京:机械工业出版社,2010.
[7] 徐存善. 机电专业英语 [M]. 2版. 北京:机械工业出版社,2012.
[8] 杨承先,等. 现代机电专业英语 [M]. 2版. 北京:机械工业出版社,2012.
[9] 马佐贤,戴金茂. 电子信息类专业英语 [M]. 北京:化学工业出版社,2011.
[10] 庄朝蓉. 电子信息专业英语 [M]. 北京:北京邮电大学出版社,2013.
[11] 闫鑫. 楼宇智能化专业英语 [M]. 北京:化学工业出版社,2009.

参考文献

[1] 刘鸿文. 等. 工程力学简明教程[M]. 4版. 北京: 人民教育出版社, 2014.
[2] 陈国良. 电子信息类专业物理[M]. 上海: 同济出版社, 2009.
[3] 王森. 机电设备检修技术手册[M]. 2版. 北京: 电子工业出版社, 2015.
[4] 张景异. 机电设备安装[M]. 3版. 北京: 人民邮电出版社, 2010.
[5] 王艳, 刘军. 机械制造基础[M]. 北京: 化学工业出版社, 2014.
[6] 范云熹. 机械设计基础[M]. 北京: 机械工业出版社, 2010.
[7] 刘晋春. 机电一体化技术[M]. 2版. 北京: 机械工业出版社, 2012.
[8] 张文灿, 等. 测控技术及仪器[M]. 2版. 武汉: 华中科技大学出版社, 2012.
[9] 韩红. 机电一体化系统设计与应用[M]. 北京: 清华大学出版社, 2011.
[10] 王春燕. 电子技术与电工基础[M]. 西安: 西安电子科技大学出版社, 2013.
[11] 陈炎, 等. 工程材料及成形技术[M]. 2版. 北京: 化学工业出版社, 2009.